DEFYING GRAVITY

DEFYING GRAVITY

A LOOK AT LIFE FROM A HIGHER PLANE

KEN IBOLD

Tate Publishing & Enterprises

Defying Gravity
Copyright © 2010 by Ken Ibold. All rights reserved.

This title is also available as a Tate Out Loud product. Visit www.tatepublishing.com for more information.

No part of this publication may be reproduced, stored in a retrieval system or transmitted in any way by any means, electronic, mechanical, photocopy, recording or otherwise without the prior permission of the author except as provided by USA copyright law.

The opinions expressed by the author are not necessarily those of Tate Publishing, LLC.

Published by Tate Publishing & Enterprises, LLC
127 E. Trade Center Terrace | Mustang, Oklahoma 73064 USA
1.888.361.9473 | www.tatepublishing.com

Book design copyright © 2010 by Tate Publishing, LLC. All rights reserved.
Cover design by Kellie Southerland
Interior design by Nathan Harmony

Published in the United States of America
ISBN: 978-1-61663-877-1
Transportation: Aviation: General
10.05.13

ACKNOWLEDGEMENTS

This book would not have been possible without the encouragement and assistance of that huge, but intimate, crowd of people we call pilots. Among this fraternity of like-minded souls, I would like to specifically cite (in alphabetical order) Dr. Bruce Chien, Chip Gibbons, Ron Levy, John Lowery, Diana Richards, and Patrick Veillette. Without them, I would know less about flying and less about myself. Many thanks to Steve Brady for giving me knowledge and skills that saved my skin more than once.

Finally, I would like to thank my family and especially my wife, Catherine, for believing in me and accepting my flying addiction with minimal complaint. Certainly without her, you would not be reading this.

TABLE OF CONTENTS

Prologue	13
Part I—In the Air	15
The Forgotten Road	17
Hearing Loss	19
Attention Deficits	21
Reality Check	23
Chinks in the Armor	25
Assistant in the Tower	27
A Second (and Third) Opinion	29
Sensible Alternatives I	31
Details of Flying	35
Catastrophe Averted	39
Playing Your Part	41
Trust but Verify	43
Back in the Saddle	47
Thanks All Around	49

Wishing You Were in the Air	53
Eyes of the Blind	55
Lights Come and Go	59
No Good Deed	61
Wrong Place, Wrong Time	63
Lost Soul	65
Rear Window	67
Great White Hope	71
The Other Ten Percent	75
Sensible Alternatives	77
Round, Round, Round We Go	79
Control Freak	81
Threading the Needle	83
Two Pilots, No Captain	87
The Best-Laid Plans	91
Part II—The Sanity of Flight	**95**
Are You Nuts?	97
The Demon in Chains	99
Greek Theater of Flight	101
The Four Dimensions	105
Shots in the Dark	107
Protective Custody	111
No Gear	115
The Many Faces	117

Flying at Work and Play	119
Meet George	123
Two Down, Among the Many	125
Two Times Two	127
Be It Resolved…	129
Mirror, Mirror on the Wall	133
Coveting Some Support	137
Sanity Lost	141
Sensible Alternatives II	145
Answers for Everything	147
Morbid Interest	149
The "Easy" Button	153
Risk, Practically Speaking	155
Feed Your Brain	159
Old Challenge, New Opportunity	161
Creating Used Airplanes	165
The Ruinous Road	169
The Mistakes of Others	173
Airlines vs. GA	177
Color Me Green	181
All Grown Up Now	185
Little Jets, Big Changes	189
The Training Dilemma	193
Breaking Routine	197

Something Old, Something New	201
Sensible Despite Chaos	205
A Study in Contrasts	209
Part III—Use Some Sense	**213**
Nobody's Perfect	215
Use the Force?	217
Denver's Lesson	219
Diminishing Returns	221
Self-Delusion	223
Shortcut on a Long Path	225
Dependency	227
Depressing Familiarity	229
OSH 'Til You Drop	231
Gotta Gotta Gotta… Scrub	233
Remarkable Detail	235
Where's the Risk?	239
The Prop Spins	241
Overlooked	243
Red Light	247
When "Broken" Isn't	251
The Safety Stool	255
DIY, Sort Of	259
Inquisitive Culture	263
Right Turn, Left Turn	267

Part IV—The Politics of Flying   **271**

    Setting the Tone   273

    No Gray Hair Here   277

    Repercussions   279

    Tarot Security Agency?   283

    Arresting the Jailer   287

    Insurance Rules   291

    Shades of Gray   295

PROLOGUE

Everyone's been on an airplane. Sure. But the fact is that most people have never been flying. The number of people who have strapped themselves into the cockpit of a light airplane is a pittance compared to the number who parade through airport security like cattle, doffing shoes and opening laptop cases at the insistence of uniformed rent-a-cops who masquerade as federal security officers.

Light aircraft and airliners do have some things in common, but they are far more different than they are alike. Airliners are, first and foremost, the business of transportation. Light airplanes are an extension of the person flying them. Airliners are the modern equivalent of the transcontinental trains that tamed the Old West; light planes are the cowboys' horses.

Just like the cowboys who explored the West, pilots are a breed apart from the rank and file of American society. One popular image holds pilots out as rich playboys who flit about on whims and damn the expense. Another portrays pilots as daredevils who tempt fate each time they push the throttle in and let the airplane accelerate down the runway.

The reality, of course, is slightly different. Though there

are playboys and cowboys who fly airplanes, they are no different from their land-lubbing counterparts who drive exotic sports cars or hot-rod street cycles. The broader population of pilots is a lot like everyone else but with a very important exception: an independent streak a mile wide.

What follows, then, is a journey into the pilot's mind. It decodes some of the secret handshakes, examines how pilots think as they seemingly defy gravity, and helps define the place in the world where private aviation lives.

This book is written by a pilot for pilots but also for anyone who has dared cast an eye skyward at the drone of an overhead propeller. Those who have, at one point or another, found themselves close to one of the 600,000 Americans who hold a pilot certificate might also find some measure of understanding of what Sergei Sikorsky, son of helicopter pioneer Igor Sikorsky, has called "a very pleasant mental disease."

PART I
IN THE AIR

THE FORGOTTEN ROAD

A world of contradictions lay beneath my feet, which were buzzing slightly with the even hum of the engine.

The Florida summer sky, so often drenched with cumulus-feeding moisture, was smooth and clear. Only a few small dots of cloud muddied the far horizon on one side. The air at 4,500 feet was dry and 71 degrees. Jimmy Buffet sang a ballad crisply in the stereo headset that also served as my umbilical to the outside world. The radio popped only occasionally as I monitored the emergency frequency. If this were anything but an airplane, there would be a beer commercial here somewhere.

Stretching below me, thousands of people went about their daily lives. I was too, of course, but I couldn't stifle a slight smirk as I wondered at the choice hand life had dealt me this day. This was not slipping surly bonds; it was ignoring them, rendering them moot, daring them to try to reclaim me as their prize.

Where others might have seen boredom or some waste of time cruising at 100 knots, I found freedom and tranquility. All too soon, I saw my destination begin to take shape in the distance. It was still fifteen miles away, but already I could make out the runway configuration at the uncontrolled field.

I dialed in the common traffic frequency, reduced power slightly, and began a leisurely descent. Through checklists and a quick verification of the airport layout, I monitored the traffic: a couple of Skyhawks, an experimental and a Commander were reporting circuits or inbound. I announced my position and called inbound as well, maneuvering to enter a left downwind for the runway the wind apparently favored.

My pattern was tight, the turns crisp; I came down the final approach course within two knots of my target speed and rolled the main wheels on the ground with not even a chirp. Stick forward as the speed bled off, then I pinned the tailwheel, exited the runway, and parked at the FBO.

It was then, standing in the shade of the wing, listening to the ticking of cooling metal under the cowl, that I realized the price of my reverie.

I didn't know the wind speed or direction, even though my landing had been perfect. I didn't know the altimeter setting. I knew only myself, my airplane, and the tinny voices that chattered in my headset.

Some may argue that old-fashioned stick-and-rudder flying on a flawless day needs not be bothered with the artificial minutia of modern regulations and technology, and they would be right. It's certainly true that on this flight, my oversight in not listening to the automated weather or checking the windsock carried no foul.

But still, my mood deflated as I realized that, for an hour in the sky, I had let the elements rule me, rather than me exercising power over the air. And I wondered if somehow, somewhere, those surly bonds were plotting to get me back.

HEARING LOSS

Pilots feel pretty smug about themselves sometimes. When they're masters of their machines and their sense of proficiency meshes perfectly with the proper caution, flying seems far from dangerous. It seems...ordinary. But there are times when even the best pilots in the most familiar environments can find unexpected challenges when they least expect it.

Recently I was on board a single being flown by two very experienced pilots. Through a bit of test flying, some practice approaches, and some touch and goes, it was obvious to me that these two veterans had the technique down cold. That thought, like sitting down while waiting for an elevator, only served to invite trouble.

The tower controller was very busy, and his rapid-fire instructions became punctuated with severe static that had me wondering for a few minutes why the new Bose X headset I was wearing carried such a hefty price tag. At first we thought it was a stuck mic aboard some noisy plane battering us with its decibels. The din was so great that, at one point, the controller transmitted our call sign three times before any of us heard it.

By the time we answered, the traffic he was calling about was about to pass behind us at a right angle. The Cessna came within about 1,000 feet horizontally and 200 feet vertically. At the time we were inbound on an ILS, although we were operating VFR, and the approach had the attention of the pilots up front.

During the course of the approach, I could sense the controller's frustration mount as, time and again, he would have to call aircraft numerous times before they would respond. In busy airspace, the conditions for disaster were right.

We missed the approach and left the area. We headed to another airport, where communications were clear. After doing some instrument work there, we headed back to the scene of the near-miss. Again, as we approached, the static was intense. Finally, we asked the controller if someone had a stuck mic, or if our radios might be malfunctioning.

"Neither one. There's something on the final approach course that's giving off that static. It's been that way for a couple of weeks. We've got lots of people looking for the problem, but so far no luck," he replied.

Here was a known radio problem at a very busy reliever airport adjacent to Class B airspace, and yet it was not Notam'd or otherwise dealt with on any official level.

It should serve as a reminder for those pilots who comfortably operate within the system that many problems have nothing to do with you, your skills, or your airplane. Fly as if there's some vast conspiracy trying to bring you down, and your approach to flying will be, well, unimpeachable. Get too smug, and an inopportune patch of bad luck may throw some static into your plans.

ATTENTION DEFICITS

I was level at 9,000 feet, hand flying through moderate turbulence, when the GPS in the panel decided to give up the ghost. No warnings, no flags; it just went black.

I was on an airway at the time and had the VORs tuned, so there was no navigational problem, but if nothing else, the GPS conveniently gives an instant answer when a voice from the back asks, "How much longer?" As I looked over to troubleshoot the unit, I got too engrossed in following the error messages the unit was spitting out on rebooting it. When I looked back at the instruments, I was banked 30 degrees and had just begun an 800-foot-per-minute descent.

With the turbulence, I didn't want to turn control over to the autopilot, so the GPS would just have to wait. It was a solid reminder, however, of the way distractions can cause trouble if they're anything other than momentary.

The incident reminded me of a trip several years earlier. My copilot was my son, three years old at the time. The gear had barely been retracted, and we were climbing IFR over the top of a busy Class B airport when he uttered those fateful toddler words: "I got to go potty."

My initial reaction, of course, was to ignore him,

because I knew he had just used the facilities at the airport not fifteen minutes before. Besides, I was much too busy to deal with it and wasn't about to cancel my clearance and turn around. In a few moments, there were tears streaming down his face, and he had a death grip on his crotch.

Right about that time we reached cruising altitude and were cleared on course. The airplane we were in had a squirrelly autopilot and a rigging problem that preventing trimming it reliably for straight and level. I was forced to divide my attention between flying the airplane and helping my son take care of business into an airsickness bag.

Although the altitude diversions that resulted never raised the ire of ATC, the attitude diversions made the next few minutes seem adventuresome.

Fast-forward thirty minutes. The little guy's nervous bladder again makes the call. Surely he's just looking for attention, I told myself. But who can ignore the pleadings of one so young? Out came the bag again. This time we did a little better.

Both incidents serve as reminders that distractions can come from anywhere. And if they only happen when you're expecting them (or when they're convenient), then they're not really distractions at all.

Have your priorities in mind at all times. Keep your skills honed enough that it doesn't take all your concentration to maintain some semblance of a heading or altitude. And if you have an autopilot installed, for heaven's sake make sure it works and you know how to use it.

REALITY CHECK

Everyone has their own little routine for making sure they use the correct procedures in the cockpit. It starts with checklists vs. flow checks and continues on to CIGAR and first power reduction after takeoff. Throughout every flight, you've got your way of doing things.

Although some people may disagree on the details of the procedures, the purpose is the same, and they all seem to get the job done. The nuances are just for hangar talk, anyway.

Sometimes, however, a pilot will zig when he or she should have zagged, and something gets broken. Identifying the source of the problem is the challenge, but there are several likely suspects.

The first is distraction. Who hasn't been interrupted during preflight and then forgotten where they stopped? How many have picked up where they thought they left off, only to find out later the nose wheel was still chocked or the baggage door was not secure? Although checklists may help, you can lose your place on a list just as easily as you can on a walkaround.

The important thing is to pick up with something

you know you've done, rather than something you don't remember doing, and take it from there.

In flight, however, things can get a little dicier.

Identifying what you need to do and verifying that the action you're about to take is, in fact, the correct one takes only a second, but that second can save mountains of grief.

One pilot I know of was landing a taildragger and got a bit of a bounce. In an attempt to go around, the pilot pushed the bottom lever on the left sidewall of the airplane instead of the top lever. Instead of advancing the throttle to full, the unlucky aviator applied full nose down trim. The resulting nose-over was predictable.

Use all your senses too. Is that handle you're about to throw round like a wheel or flat like a wing flap? Your hand can tell you even if your eyes are occupied elsewhere.

The idiosyncracies of flying are all around, so keep an eye out for them every time you fly. And fly with other pilots any time you can. A low-time pilot can learn from a high-time pilot, yes, but the reverse is also true.

Notice the care a new pilot takes with the preflight inspection, or how picky he is on the mag check. Have you blown through those lately?

Pilots with equal logbooks can teach each other something, too, because everyone has had unique experiences that can be a lesson to others. And maybe you'll learn a neat trick, like "lights, camera, action" for strobes, transponder, and fuel pump.

If so, pass it on.

CHINKS IN THE ARMOR

I had strapped into my Lance for the semi-annual recurrent instrument training, which I take every six months, regardless of my perceived proficiency or legal currency. For more than two hours, flight instructor extraordinaire Steve Brady had me put the airplane through its paces while he put my brain through paces of its own.

He kept looking for chinks in the armor, while I was determined to show as few as possible.

Steve is one of those flight instructors who knows that his job is more than pencil-whipping a log book. He probes for weaknesses and, once he finds them, assaults them mercilessly. As a consequence, flying with him is always demanding, exhausting, and extremely productive.

This session was no different, including about 40 miles of DME arc to within 0.1 mile and some dreaded NDB work in strong, gusty winds. After a couple of hours, I was happy to return to the airport. But the fun for the day wasn't over. We parked the Lance and fired up the Citabria for some tailwheel work with the strong crosswind.

The Lance is a fairly big single, and the airplane responds slowly to control inputs and power changes. You need to think ahead and anticipate how the airplane will move. The Citabria is the opposite. Light and nimble, you think about what to do, and the airplane almost reads your mind.

The stability of the Lance and the squirreliness of the Citabria in the crosswind made for an interesting transition that day, but I promised myself the little taildragger would not get the better of me.

After almost another two hours of being victimized by the sadistic instructor, I was ready to go home. Unfortunately, we were still 900 feet above the runway with another of those potential ground loops waiting. That's when I did something stupid—or rather, didn't do something smart.

I should have confessed my fatigue, turned the airplane over to the rear-seater, and had him take us back to the airport. Instead, I pressed on, my reaction time dangerously slow. I sort of pointed the airplane at the runway and thought, *Come on, Trigger, take me home.*

A tailwheel airplane in a crosswind, of course, is not a trusted horse. The landing that ensued didn't bend anything, but it sure wasn't pretty. My best lesson of the day turned out to be the last one.

Fatigue is an insidious foe that creeps up on you. Usually you know it's there, but somehow you just don't care.

The problem seems to be worse in the winter, when more flights end after dark, and the darkness amplifies the fatigue factor. It's no coincidence that loss-of-control accidents dominate the winter months.

Stay sharp, and put the airplane on the ground before you're too tired to care how it gets there. Once you push it too far, you'll be lucky if your arrival is only embarrassing.

ASSISTANT IN THE TOWER

Another pilot and I were doing pattern work at a controlled airport. It was a weekday, and there was only one other airplane, a Cessna 152, in the pattern with us. It didn't take long for us to start wondering what we were doing there.

The other airplane was flying left-hand patterns that would have accommodated a B-52. The downwind leg was fully two miles from the runway. By the beginning of the base leg, the Cessna was at full flaps and lumbering along toward a two-mile final. The controller didn't seem to notice.

After following this airplane for two patterns, we told the controller we were going to fly a tighter pattern for safety, and that we would make minimum speed to avoid overtaking. We hoped the other pilot would get the hint.

He didn't, but the controller sure did. After our next takeoff, we were cleared to remain in right closed traffic and, for the rest of the hour, completed two trips around the pattern for every one the Cessna made. Only once did our traffic patterns conflict. Guess who got the 360. Wasn't us.

That was one case where a heads-up controller made our day. Turns out, there was an instructor and a student pilot on board the Cessna. Regardless of whether the instructor had financial motives for flying the wide pattern, he was not doing his student any favors—either from a financial standpoint or a safety standpoint. Flying a wider pattern may slow things down for the student, but if the need is so great, eschew the touch-and-goes and conduct full-stop landings followed by a debrief and taxi back.

Another flight also highlighted a serious safety issue in some parts of the country. We were leaving our home base, Orlando Executive, where a new flight school had recently opened with a large roster of foreign students.

On this particular flight, we were taxiing to the runway just as a solo student pilot came on ground frequency. While I believe in cutting students as much slack as possible (because we've all been there), this particular man's command of English was so poor his communications were unintelligible. After asking him several times to repeat himself, the frustrated controller finally resorted to having him spell everything he was saying.

Think about how long it takes to say "charlie echo sierra sierra november alpha one seven two" and imagine what that did to the traffic flow. While in this case it didn't affect us because we were on the ground, this pilot would soon be in the air, going to other airspace, and wreaking all sorts of unknown havoc.

These two incidents underscore the fact that you are not excused from your responsibility to be alert and vigilant just because there's someone on the ground giving you occasional instructions over the radio.

A SECOND (AND THIRD) OPINION

Pilot 1 is doing a run-up and notices an excessive magneto drop on one side. He runs it up to nearly full power briefly, the engine smoothes out, and the mag drops become equal and normal. He takes off and makes an uneventful flight. When he returns, he tells Pilot 2 of the problem and the solution.

The next day, Pilot 2 is doing a run-up on the same airplane. He notices an excessive mag drop on one side, and recalls Pilot 1's solution. Recalling that full power solved the problem for Pilot 1. He takes the runway and advances the power to full. The plane takes off as usual, but the com radios are filled with ignition noise. At an intermediate destination, a mag check shows equal drops that are on the far end of the allowable range, but within limits. Pilot 2 returns safely and tells Pilots 1 and 3 about the experience.

A mechanic is summoned. After investigation, he determines that the ignition harness needs to be replaced. The part is ordered and installed. Pilot 3 gets the duty of testing the repair.

Pilot 3 notes that the com is free from ignition noise

and taxis to the run-up area for a test flight. At run-up, the mags show an equal drop, but the drop is excessive and the engine runs poorly. Recalling Pilot 2's experience, he takes the runway and advances the power to full. The engine refuses to develop more than 2,000 rpm. Pilot 3 aborts the takeoff and taxis back to maintenance. The new ignition harness is found to have a flaw that allows several of the leads to arc.

Troubleshooting problems in aircraft is complicated when more than one pilot flies the plane. Describing the nature of the problem to other pilots can lead to a distorted impression and risky behavior. If the message must then be relayed to a mechanic, the chances of a misdiagnosis increase even more.

In the true story above, all three pilots believe they acted cautiously and responsibly, especially given the 6,000-foot runway available. The important thing to remember is that not all mechanical failures involve components working fine one minute and not working at all the next. In this case, the ignition harness had hidden corrosion that gradually made performance decline.

It illustrates the necessity to keep careful track of the airplane's performance and note even small changes over time. This will not only help every pilot who flies the airplane make an individual risk assessment but will help the mechanic pinpoint the problem at whatever point maintenance is done. That will not only save you money, but it could also save your life.

SENSIBLE ALTERNATIVES I

Sometimes you recognize get-home-itis when you've got it. At that point, you have to step back and make sure you're making the right decision because the facts support it, rather than because your emotions have taken hold.

During a recent weather briefing for a flight from Wilmington, N.C., to Orlando, the disease hit me like a rogue wave crashing over the bow of a boat. Sunset was near and a line of strong thunderstorms was moving in from the west. From the looks of the radar in the FBO and the view out the window, I estimated we had about twenty minutes to get off the ground or we'd be stranded overnight.

We were prepared to stay, but it had been a long day for the kids, and I knew they'd sleep the whole way if we left right away. If we left in the morning, they'd be fresh and rested and ready for anything other than sitting in the cabin of the Lance for three hours.

The briefer painted a pessimistic picture of the route home. Thunderstorms were predicted inland in South Carolina and Georgia. Central Florida had current thunderstorm activity showing on the radar. As I continued my briefing, his tone made it clear he thought I should abort the trip. When I filed my IFR flight plan, the disapproval in his voice was noticeable.

I recognized my eagerness to depart, but when I stepped back and looked at the dynamics of the flight, it didn't seem all that bad. The question I had to ask was whether I was merely rationalizing my decision to go.

Although the pressure was building to get the prop turning, I took a few moments to think about what was happening. The storms were moving in, but our point of departure was near the south end of the line. If we could get out before they arrived, we'd easily be out of their reach.

The same front was forecast to give birth to storms in South Carolina and Georgia, but nothing had erupted yet and those forecasts were inland; we'd be flying along the coast. Our route kept us a few miles offshore most of the way, where the air would not be as unstable.

In central Florida, the existing storms were the isolated late-afternoon variety that occurs almost every day, and I knew they'd dissipate by the time we arrived at about ten thirty at night. If they didn't, I was confident I'd be able to pick my way through them with the help of the Stormscope.

The routing was simple: direct Savannah, direct Orlando. There was no traffic at the field, and I was immediately cleared to take off and climb to altitude. We watched a beautiful sunset over the clouds as the towering storms moved in slowly behind us and to the west.

I suppose the briefer thought he was doing me a favor by cautioning me repeatedly about the weather, but I also thought—then and now—that he was being overly conservative and that I made the right call. There was enough margin that even if the weather was worse than forecast, I still had plenty of options, including flying to the Bahamas and waiting it out.

Come to think of it, that's probably the toughest call I made all day.

DETAILS OF FLYING

Over the course of a pilot's career, a glance backward in the logbook will show more than a snapshot of the pilot's experience. It will also show the attitude the pilot has toward flying. A logbook filled with nothing but hours in a Cessna 172 reflects a very different pilot than one who has only a few hours in a variety of seaplanes, aerobatic planes, piston twins, and a double handful of high performance single-engine models.

That's not to say one is necessarily better than the other, because each pilot has learned some essential lessons that have escaped the other.

In my twenty-five plus years of flying, I've owned four different airplane types (Mooney 201, Piper Lance, American Champion 7GCAA, and Pitts S-2A) and flown more than fifty different models, including seaplanes, warbirds, light-sport aircraft, turboprops and light jets.

When you only fly one airplane, as most owners do, the airplane and its performance become as familiar to you as those spots you have to shave around. You know the airplane's every squeak and quirk. You know exactly how much to change the trim when the gear goes out or the flaps come up. You know its fuel burn almost to the ounce and its speed to the fraction of a knot. More importantly, you can just tell if something's not working right.

On the other side of the coin, flying many different airplanes, as renters typically will, means you never take performance for granted. Your scan of the instruments is a bit more vigilant, and you are on the alert for such things as checking the brakes before you taxi or a hint of a stall when turning base to final.

The true dynamic has been repeated several times for me. Because of the aircraft flight test reports I've written from time to time, it's happened more than once that I've flown somewhere in my airplane, tested another airplane, and then flown home. On one trip, I flew a Micco SP20 at the company's factory in Fort Pierce, Fla. The flight down in the Lance was a snap and, after a factory tour and some hangar flying with company personnel, we mounted the Micco for a demo flight that explored the flight envelope. While we didn't really wring it out, virtually every second of the flight was a test of some aspect of the airplane's handling. By the time we landed, not that many ticks of the Hobbs had gone by, but I was a different pilot from when I started.

My senses were primed to pick up the slightest nuance, true, but my alertness was also a bit done in by the need to focus attention so continuously. After only a short break, it was back to the Lance for the trip home.

I could swear I was flying a different airplane. The trip south had been of the point-and-go variety, while on the trip home I noted every burble of air slipping past the Piper's stubby wings. One burble in particular got my attention, and it turned out to be a gear door that was not rigged correctly.

Fly an airplane you know, but sometimes, fly one you don't. It may show you something you've been missing all along.

CATASTROPHE AVERTED

On a gorgeous VFR day, the occupants of two airplanes never knew how close they came to having their days ruined. I watched it happen.

It was time for my flight review, so my instructor and I loaded up the Citabria (sans parachutes) for some air work in the boonies west of Orlando. Then we flew north to Leesburg, an uncontrolled airport with intersecting runways. Winds were light and there was a thin layer of cumulus that started at about 2,700 feet and went up to about 4,000.

Traffic was using runway 13, and there are lakes at both ends of the runway. It's usually a quiet little airport, although occasionally there are trainers around shooting touch-and-goes or NDB approaches.

We were in the pattern with three other airplanes. One was practicing landings, and the other was heading for a full stop. Also operating nearby, according to the CTAF, was an airplane practicing the NDB approach, and another practicing holds over the NDB at 4,000 feet.

It was the kind of day when everyone was using perfect manners. Calls were clean, courtesy apparent, and eyeballs were open.

As the Apache on the NDB approach neared the airport, he was rapidly overtaking me. We agreed on what each of us would do, and I turned my attention to the Cessna I was following, which was turning from left downwind to base.

A Learjet eight miles out called for a straight-in approach to 13. The pilot said "any traffic in the pattern, please advise." The Cessna ahead of me called in. As did I, despite my feelings about the phrase "any traffic in the area, please advise."

Suddenly a medium twin appeared on short final, apparently having flown a close base under the traffic on downwind. As he crossed the edge of the lake at about fifty feet agl, I could see the Lear halfway across the lake, closing the distance fast between himself and the mystery twin.

The twin cleared the runway before the Lear touched down. There was no immediate sigh of relief and no "ohmygoddidyouseethat?" My instructor and I looked at each other (as well as you can in a tandem-seat airplane) and wondered aloud about how much it would have taken to write a very different end to the story.

The twin's lack of communication was troubling, as it obviously was not the kind of airplane that goes NORDO too often. The Lear's decision to fly a straight-in with multiple airplanes in the pattern is perhaps typical but not very welcome on a nice VFR day.

The Aztec flying the NDB used the radio, but his low approach could have spelled disaster if the Lear had gone around because of the twin on the runway.

Sometimes we wonder why non-towered fields are called "uncontrolled." Other times it's readily apparent.

PLAYING YOUR PART

It's easy to feel like you have a guardian angel when you're flying into an airport with a functioning control tower, especially one that's equipped with D-BRITE and well-trained controllers. But there still is no substitute for peeled eyeballs and an awareness of what other airplanes are actually doing (as opposed to what they're supposed to be doing).

Recently I was nearing a busy controlled airport just as controllers were switching from one runway to the opposite one. There were multiple airplanes on instrument approaches, several waiting to take off, a couple in the pattern, and a few more approaching pattern entry. I was cleared to enter a right base for the new runway, but by the time I got close, traffic in the pattern and on the ground prevented such an easy entry.

I was told to follow a Cessna on right downwind. I had the traffic in sight and slipped in behind it. A few airplanes finished their instrument approaches to the old runway, making low approaches. As the Cessna in front of me turned final, I started my base leg. Tower asked me to extend downwind a bit, giving me vectors that took me away from the airport about halfway between a downwind heading and a base leg heading.

Just as I reached the extended runway centerline, I was told to turn final. I was about two miles out and just below pattern altitude. The controller threw me another curve ball. I was to offset my final approach to the left to let a Mooney take off in the opposite direction. No problem.

The Mooney pilot was instructed to make an immediate takeoff, warned of traffic on a two-mile final, and told to make a left turn to a heading of 360 as soon as possible after takeoff.

I began counting one Mississippi... two Mississippi... Thirty seconds later, the Mooney took the runway. I was a mile out. The Mooney rotated. The gear came up. The airplane stayed on runway heading. I was less than a half-mile out, lower than the Mooney, and engaging in slow flight. I was about to call for a go-around when the airplane passed abeam my position. I pushed the nose down to increase speed, cranked in two 90-degree turns to align myself with the runway, and landed.

When the Mooney finally climbed above 700 feet, it made a lazy turn toward the north but stopped short of its assigned heading. The controller came on frequency with a tone like a scolding parent. "Mooney 123, turn left immediately heading 360. That's 360. North. Traffic, twelve o'clock, opposite direction, less than a mile."

A change of runways when traffic is busy is a complicated mission for ATC. The Mooney pilot's head wasn't in the game. The controller considerately allowed the Mooney to take off into traffic instead of taxiing to the other end of the runway. The pilot should have responded by listening to instructions and promptly executing them.

ATC and PIC blend into a partnership. When ATC does its part, make sure you do yours.

TRUST BUT VERIFY

It was time to introduce a non-pilot friend to aerobatics. He had been expressing interest for some time, and finally our schedules meshed, and we headed out to the airport to strap on some unusual attitudes.

He had flown in the Citabria once before, on a flight where we did some spirited maneuvering but no aerobatics. I like to introduce neophytes this way because it allows them to get used to the motion, and I can avoid having to wash out the interior of the airplane afterward.

Before that first flight, we were waiting for the fuel truck, and I asked him how much he weighed. "A buck eighty," he responded. I ran the weight and balance and found we could carry full fuel and still be comfortably within limits.

On the next flight a few weeks later, I dipped the tanks and found twenty gallons. I ran his weight, his parachute, myself, and my chute. We were slightly into the normal category, but by the time we burned off some fuel getting to the practice area, we'd be right within the aft limit of the aerobatic envelope.

We cleared the area, and my passenger asked for a loop.

Because it was his first trip, I decided to do one maneuver at a time and give him a chance to stop me if he started feeling uncomfortable.

The first one was a little sloppy, which I attributed to the relatively light 2-g entry pull. I cranked in just over 3 g's on the next entry, and it really fell apart. The airplane bucked and buffeted, and it was difficult to complete the maneuver with any semblance of order. It had been two weeks since I'd flown any loops, but I had a hard time believing my technique had deteriorated so badly.

The third entry was clean, the pull brisk, and the loop was clean until we were about thirty degrees short of inverted. At that point the airplane decided to act up again, with the stall horn bleating and an unusual airframe buffeting I'd never felt before.

Duh. At long last the light in my brain clicked on.

I pointed to some isolated storms north of our position and said I didn't like the way they were moving in. I cut the flight short.

As we sat around the lounge afterward, I made some self-deprecating comment about how sloppy the maneuvers had been. As if on cue, my friend said, "Yeah, well, last time we flew and you asked me what I weighed I just guessed. I think I told you about fifteen pounds less than what my scale says."

After he left, I re-ran the weight and balance using his actual weight. We had been behind the aft limit for aerobatics but well within the normal category.

Asking a passenger's weight (or asking to weigh the baggage) can be embarrassing. They may flatter themselves with their answer. I was lucky this time and learned a valuable lesson: Trust, but verify.

How to do that effectively and discreetly is a little harder. I liked the strategy DayJet reportedly used on its short-lived air taxi service. A scale built into the floor would weigh people as they approached the check-in desk, surreptitiously showing the weight of the passengers and their bags to the agent.

Or so I've been told.

BACK IN THE SADDLE

Back in those uncertain days immediately after 9/11, the nation's airspace was limited to aircraft flying on IFR flight plans. A few days later, VFR flight was permitted but only under some circumstances. At the time, my airplane was based at Orlando Executive, which was part of what the government called "enhanced Class B" airspace—no VFR flights allowed, except training flights with an instructor.

When that ridiculous limitation was finally lifted, my immediate response was to go for a fun flight. Freed at last from its hangar, my Citabria virtually leaped into the sky despite a load that was nearly at max gross weight.

We reached 500 feet well before the Cessna 172 that had taken off from runway 7 in front of us, and we were cleared to make an early left turn for a northwest departure. This was a far cry from the almost iron-fisted display of IFR flying that had been the order of the day for nearly six weeks, adding to the sense of freedom. Tapping into the karma, I slowed the airplane and opened the window. Was it noisy? Yeah, but worth it.

Once clear of the ECB, we undertook some moderately aggressive aerobatics. My passenger, also a pilot, didn't have the stomach for much, so we diverted to a nearby uncontrolled airport for some pattern work.

He flew the first pattern, a stately affair more in line with the Cessna 172 he normally flies. It was clean and competent, with a decent landing even though he hadn't flown a tailwheel lately. Then it was my turn. I turned it up a notch, staying high and tight in the deserted pattern, then aggressively banked and slipped to the threshold.

I think he got the idea.

On his next pattern, his turn to the runway was nearly a wingover. I don't know whether the excessive bank was due to the fact that he wasn't familiar with the sporty handling or whether we had unknowingly entered into a game of "Hey, watch this!"

I wasn't much in the mood to find out and was somewhat relieved when two other airplanes showed up to put an end to the more aggressive part of our game. We both flew a few more sedate patterns, with wheel landings and three-points, and then we moved to a nearby grass strip for a taste of nostalgia. I operate mostly from pavement, and I'm always amazed when I'm reminded how nicely airplanes land on a well-maintained, dry grass strip.

All too soon, it was time to go home. I left the airport with a satisfied smile, glad to be free to fly on my own terms instead of someone else's. But at the same time, that brief bout of one-upmanship gnawed a bit. We nipped it in the bud this time, but I can't help but think of the variety of situations in which such a momentary lapse of reason might be all it would take to find grass in the windshield rather than in the tailwheel springs at the end of the flight.

THANKS ALL AROUND

A pilot I know usually starts (and often completes) his preflight in the car on the way to the airport. He calls Flight Service from the cell phone, takes notes if he can, and files his flight plan. Once at the airport, he pulls the airplane out of the T-hangar, makes sure it's got both wings and still has a propeller, parks his car inside, gets in the bird, and away he goes.

Most pilots are a bit more diligent, calling Flight Service before leaving home or the office and conducting a more thorough walk-around before parking the car in the hangar and heading out. Some are downright anal-retentive about it, plotting the weather for days before a flight and conducting a 100-hour inspection before lighting the fires to defy gravity.

One year, we were flying out of Orlando for Atlanta in a Turbo Stationair for the Thanksgiving weekend. The kids wanted to see their grandparents and I wanted to see my grandmother, who, at ninety-four, wouldn't have many Thanksgivings left to share.

The trip north was uneventful, as a high-pressure system dominated much of the route of flight. A few clouds along the way and relatively light winds at 11,000 feet brought us into the Atlanta area, where some low clouds had developed. But an ILS into DeKalb-Peachtree broke me into the clear 500 feet above minimums.

The holiday was as holidays are, including that sprinkling of sideways glances from seldom-seen relatives who still find it hard to believe some people actually fly their kids hundreds of miles at a stretch in light planes. All too soon, Saturday arrived and it was time to pack up for home.

Yeah, Saturday.

When making these personal journeys, especially with kids, I like to leave a day at the end of the trip. Ideally, that gives everyone a day to relax before heading out for school and work on Monday. It also leaves time to deal with weather, airplane problems, and other unexpected delays.

This particular Saturday was rainy and messy in Atlanta, but ceilings were reasonable throughout central Georgia, and south Georgia and Florida were forecast to be good VFR weather. The briefer said conditions around Atlanta were likely to improve as the afternoon wore on, but I wanted to get home by dark to try to keep the kids' sleeping schedules somewhat intact.

We skated along the top of a layer of cumulus, punching through the occasional top but mostly skimming about 50 feet above the clouds. I kept looking behind us to try to catch a glimpse of our wake vortices. The OAT read 2C, so I kept an eye out for ice, but all I got was a little water on the airframe.

A controller idly commented on how quiet the frequency was, bemoaning how busy it was likely to be the next day as more people rushed to get home.

Then I glanced over at the Stormscope. Behind us, about 100 miles to the northwest, stood a wall of dots. Not radial spread from a strong storm, but a huge band of serious weather. As we tooled along, we gradually outran the front. By the time we got to the Florida state line it was a distant memory.

When I got home, the evening news told tales of tornadoes and severe weather spawned by the front.

If departure had been delayed an hour, I would have smacked into the front. With only 145 knots of cruise speed available, making an end-around would have taken forever. That hour would have cost me the day.

In retrospect, I realized my preflight planning had effectively ended when the briefer forecast predicted improving weather through the afternoon. The front that moved in behind me had been visible on the "regular" weather report the day before, but it exceeded expectations in how quickly it moved east.

I was glad I had that extra day, but even happier that I didn't have to use it.

WISHING YOU WERE IN THE AIR

Many wags love to toss around clichés as if they're the clever soul who invented them. How do you know there's a pilot in the room? He'll tell you.

There's a grain of truth to most of them, of course, but gosh, it gets tiring explaining the mystery and the dynamic environment of flight with such cookie-cutter phrases.

While we all might dust off a cliché from time to time, there are times when you might live one and wish you hadn't. That's when it becomes crystal clear just why the cliché has emerged from the fermenting stew of language.

However, recently I turned a cliché on its tail and discovered a kernel of wisdom there as well. "It's better to be on the ground wishing you were in the air than in the air and wishing you were on the ground." Now typically that's used as a wrap-up for someone who has scared the dickens out of himself (or herself) by flying through bad weather or with mechanical problems that should have kept them on the ground. For me, it wasn't quite that way.

I'd been battling a head cold for several weeks, and

it spawned a full-blown ear infection. One ear sounded underwater and the other one was well on its way there.

I had a flight that I reckoned I just had to make. I knew it would be iffy. I even joked about it to the right-seater as we climbed into the Seneca. I figured it'd be best to stay at a fairly low altitude and make gradual descents. Been there; done that.

To make a long story short, I was wrong. It was a bad call.

By the time we climbed to 4,500 feet, the pressure in my ears was nearly bad enough to make me turn over the airplane to the other pilot. Then a few swallows put things right. Close enough, anyway.

Coming down was another story. After we landed, the controller's voice in my ear was agony, even though the volume control on my headset was much lower than normal. Every head movement made me feel like a particularly unpopular resident of a CIA secret prison.

Did I say I was wrong?

Conversation on the ground as we tied down the airplane was forced. I guess I talked to him, but I don't have a clue what he was saying. I wallowed in my misery. I was on the ground, and I really, really wanted to be back in the air. Maybe another flight with a 50 fpm descent would do it. I gave passing thought to giving it a shot.

But this story has a happy ending. As I was walking to my car, I got a squealing earful that equalized the pressure enough to restore my self-esteem. And for the first time in a week, my ears came up from underwater.

I drove straight to the doctor for some long-overdue antibiotics.

THE EYES OF THE BLIND

Early on a beautiful February morning, I departed on a 370 nm flight from Orlando to Atlanta in a VFR Citabria. Winds aloft were light, and the air was smooth, but the tranquility was battered at my mid-flight fuel stop.

As I rounded out to settle into the three-point sweet spot for landing, the aircraft unexpectedly bounced. I reined it in, a bit flustered by the momentary loss of control, and taxied to the ramp.

Later, when landing at my destination, the approach was only slightly complicated by a mild crosswind. Again, as I started to settle into the three-point attitude, the airplane took an unexpected hop. Because I was a bit tired after four hours of hand-flying, I wasn't so quick to respond, and the airplane began what seemed like a slow-motion trip toward the right side of the 100-foot-wide runway.

Finally, I regained control and taxied red-faced to the FBO.

A few hours later, while riding as a passenger in a car, I unraveled the source of my two botched landings. A week earlier I'd been forced to admit I was getting old and got glasses—my first pair since undergoing vision correction surgery seventeen years earlier. Not only that, but they were bifocals.

In the car I noticed that the sidewalks looked about two feet lower through my glasses than without them. I had simply been misjudging my altitude at the flare because of the refraction of the glasses.

The next day, I departed for home, armed with the knowledge that I'd have to "flare high." I was confronted by 30 to 50 knot headwinds for my return trip, and at my fuel stop faced 18 knots down the runway. I kept reminding myself to flare "early," and even with the strong wind landed comfortably and stopped in what seemed like 200 feet.

When I arrived home nearly six hours after beginning the flight, I was facing winds eighty degrees off the runway at 20 knots, gusting to 28. As I bounced along short final, I thought to myself, *Flare early? No way this is going to work.*

As the runway loomed nearer, I was apprehensive of my technique, especially given the previous day's performance. I considered abandoning the destination in favor of a nearby airport with the runway more aligned to the wind but decided to give myself a single chance at the crosswind landing. I resolved to be spring-loaded for a go-around and diversion if anything didn't look right, apparently ignoring my discomfort with my vision.

Ready to bolt at any second, I was able to stick the airplane to the runway with minimum bounce and almost no swerve. The fact is, the taxi to the hangar was actually more challenging than the landing. The wind was strong enough I could not turn right off the runway to taxi to the hangar; I had to turn 270 degrees left.

I had taken for granted that the new glasses would

improve my flying by increasing visual acuity. I hadn't thought of the other ramifications. A more challenging initial flight could possibly have resulted in a groundloop. Without the accidental epiphany of the change in my peripheral vision, a lousy outcome on the final leg is certain.

Sometimes, it seems, everyone needs protection from themselves.

LIGHTS COME AND GO

Night flight is one of the most beautiful times to be in a light aircraft. But there are some times when the unexpected jumps up in a way that's terrifying at best, deadly at worst.

I was flying three passengers in a VFR Archer from Fort Myers, Fla. to Orlando one evening. We departed Page Field as the sky was just losing its grip on the last golden vestiges of the setting sun. Departure had been delayed a bit while we waited for some thunderstorms north of the airport to move east and out of the way.

After takeoff, we were greeted by smooth air and a spectacular light show from the lightning that still exploded within the distant clouds. No one spoke much, enthralled as we were with the magnificent forces being unleashed.

As we approached our destination, I contacted Orlando Approach for flight following, and I hoped a shortcut through the Class B airspace. I got one out of two and, as it turned out, couldn't have been happier.

I was vectored around the west side of the Bravo airspace and instructed to descend below 1,600 feet to remain clear. As I descended, I saw the lights up ahead that clearly outlined central Florida's tourist district.

We were about two miles south of "the attractions" (as they're affectionately known by locals) when the kindhearted controller gave me a heading. As I rolled out of the turn, we got another light show. The fireworks display from Walt Disney's Magic Kingdom lit up the sky below us.

After that, it seemed like the rest of the trip would have to be anticlimactic. On a right downwind to runway 25, I reduced power and configured the little Piper for landing. When the time was right, I turned right base.

After a few seconds, I glanced out the right side window to judge the turn to final and saw absolutely nothing. For two fleeting seconds my heart was in my throat, and I had a sudden and overwhelming sensation of dizziness.

Fortunately, I quickly remembered the small lake off the approach end of 25 and realized my view out the window had been down into that lake. A switch of the head, and the runway lights welcomed me down.

The amazing part about the flight was how quickly the disorientation came when I didn't see what I expected to see—when instead I saw nothing at all. I like to think the rising panic would have been transitory even if I hadn't acquired the runway right away.

Funny, though, how a flight defined by the light shows is now a memory categorized by the darkness.

NO GOOD DEED

File this one under "Pilot as good guy, sometimes to a fault."

I like to help out controllers when I can. If the departure end of the airport is stacked with departures while I'm making a leisurely VFR approach, I'm quick to volunteer to extend downwind to let a couple of airplanes out. If I'm in a slow airplane approaching final in front of a fast one, I'll offer to take a vector or turn to put me behind the other airplane.

While there are a lot of pilots who share this attitude, some take helpfulness to counterproductive extremes.

I was transitioning some busy airspace when a pilot called the tower, reporting "inbound from the southwest." The controller asked his position relative to a prominent landmark, and the pilot said he was a mile southwest of it.

That was apparently enough for the controller to identify the airplane on the D-BRITE, but the pilot, whose radio manners suggested he doesn't get out much, insisted that he needed to ident so the controller could positively identify him. The controller said, "Thanks, but I've got you." The pilot persisted until, despite several assurances from the controller that the airplane was identified, the pilot reported identing his transponder.

No harm, no foul, but from that point on the controller's attitude toward this airplane was condescending, if not downright hostile, for the remainder of the flight.

Fast-forward another week. I was departing Orlando Executive and heard an incoming VFR airplane making initial contact with the tower. He reported "inbound over Florida State University with information Mike."

Not all controllers are whizzes at geography, but this one was sharp and knew that the pilot either was mistaken nor had an extremely powerful radio, since Florida State University is about 200 miles away in Tallahassee. He asked the pilot to say direction and distance from the airport. The pilot responded again that he was flying "right over the top now of Florida State University."

The controller, probably suspecting the pilot meant the nearby University of Central Florida, asked if the airplane was northeast of the airport. The pilot responded, "Right over the top of Florida State University."

This persisted for what constitutes an eternity in frequency time, with the pilot responding to each question on his location with a reference to the wrong location. If the controller was playing with the pilot's head, he was good at it, because finally he said, "Cessna 34A, I guess you're over some school somewhere. Just fly to the airport and let me know when you get here."

Unfortunately, it was then time for me to make a frequency change, and I didn't get to hear the rest of the flight.

Most adults have learned what these two pilots failed to recognize—help ain't help if it confounds the other person.

WRONG PLACE, WRONG TIME

The first time I tackled a real hill on snow skis, I stood at the top with a confident smile and thought, *I can do this.* Somewhere in the middle, I was riding a tiger and afraid to dismount. At the end, with shaking knees I screwed up my bravado and said, "Let's do that again."

So it was with a recent flight, except for the "do it again" part.

I was flying on a day when thunderstorms were popping up throughout the Southeast like fireflies on a summer night. Despite the best efforts of the National Weather Service and the Stormscope, I was heading right toward one as I flew toward my destination.

The approach controller had cleared me for the GPS approach and given me vectors toward the final approach course. I was in and out of cumulus, mostly in, but eventually found myself in a patch of sky where I could see a moderate buildup that appeared to be short of convective size. It was at my twelve o'clock, and I was soon in IMC again.

Just about the time I was expecting to be switched to the tower, Approach informed me his ASR-9 showed a level four between me and the airport, followed by that lovely phrase, "State intentions."

I decided to divert to a nearby airport behind me, where I knew the weather was acceptable because I'd just flown past it. He gave me a vector to take me toward the ILS, and as I turned, the cloud I was in decided it was time for puberty.

The updraft I hit carried me up, up, and away. With the power at idle, and the airspeed annoyingly close to the yellow arc, the VSI was still close to the upper peg. The airplane pitched and yawed and felt like it was working up the courage to roll, but I managed to keep wings level, more or less.

After what seemed like an eternity, but really wasn't more than fifteen or twenty seconds, I popped out of that mess and into a cloud that was merely mortal. The Stormscope was clear. I'd not seen a drop of rain. As I was getting vectored to my new destination, I kept an eye on the cloud where I'd hit the updraft. Less than ten minutes later the Stormscope lit that section of the sky.

I would like to say I shook it off, landed normally, and looked back at the airplane as I walked away, saying, "Let's do that again sometime." Not quite.

I flew a terrible ILS, chasing the needle back and forth as I focused on the past rather than the future. When I broke out, I circled to land in what was not, shall we say, a graceful arrival.

It was only then that my passenger turned and said with a restrained voice, "Well. That was a little hairy."

LOST SOUL

High ceilings early in the day had, by mid-afternoon, given way to a 3,500-foot overcast and declining visibility. Below the clouds it was still decent VFR and would remain so for several hours, after which some storms in advance of a cold front would hit the Florida peninsula. But a haze had settled that cut visibility to about seven miles.

I was flying the Citabria from a meeting in Venice, Fla., back home to Orlando at 2,000 feet, and I had tuned in the Lakeland Tower frequency in advance of transitioning that airspace. Things were pretty quiet until an uncertain voice came on the radio.

"Uh, Lakeland? This is Bonanza 12Z calling for Lakeland. Lakeland Tower."

"Bonanza 12Z, Lakeland Tower."

"Uh, I'm coming in there and, uh, I can't find, uh, I can't find the airport."

Note that Lakeland has a VOR on the field, which was in service at the time. Winds were out of the southeast, and runway 9 was in use.

The controller helped the pilot orient himself and gave him a 160-degree heading that would put him on a left base for runway 9. Over the next few minutes, however, it became apparent the pilot was either unable to hold a heading or had some kind of unspoken equipment issue. He ended up flying east. Then west. Then east again.

Despite attempts by the controller to guide him in, the pilot was instead flying back and forth looking for the airport where he thought it ought to be. In exasperation, the controller started suggesting prominent landmarks. The area around Lakeland is dotted by big lakes, one that has a large power plant on the shore. An interstate highway runs past it a few miles away.

When finally the controller had the Bonanza lined up on final for runway 9, he asked the pilot if the runway was in sight. It was not. The flight came in closer.

"Do you see that smokestack about a mile in front of you?" the controller asked. "Fly right over that. The runway is twelve o'clock and two miles."

The pilot reported seeing the smokestack, but said he still couldn't see the runway. By now, I was about six miles out and could see the Bonanza coming up on the runway.

Finally, on about a mile final, the pilot reported the runway in sight. When he was on the ground, he told the controller he'd been looking on the wrong side of the highway for the airport.

Now I don't mean to imply this pilot was irresponsible because he got disoriented or lost or whatever you want to call it. The shocking thing to me was the complete and utter lack of proficiency he showed, both in trying to follow the controller's instructions and using the radio.

Not all pilots fly as much as they need to, and not all pilots get routine practice in radio communications. If you fall into one of these categories, do something about it before your Bonanza (or whatever) gets the better of you.

REAR WINDOW

Just for the fun of it, I decided to learn to fly my taildragger from the back seat. The Citabria is outfitted to be flown solo from the front seat. There's a stick, push-to-talk switch, throttle, and carb heat in the back, but from there you can't change the radio frequency, adjust the mixture, hit the start button, or manipulate any of the panel controls. Because there has to be a person in the front, you can't see much either.

Nonetheless, I wanted the challenge, in part because I thought it might help me with the Pitts I dreamed of owning one day, so I enlisted the help of one of my favorite instructors and off we went. His first item of familiarization was to have me climb into the back, and then he lifted the tail so I could get a mental picture of the takeoff attitude, using only the wingtips as my guide. I knew this was not going to be an ordinary day.

Taxiing was more Pitts or P-51 or Stearman than Citabria, with S-turns a requirement for seeing what's in front of you. Suddenly the taxiways seemed much narrower.

The runup and mag check were normal. I dictated a pre-takeoff flow check to him, which was unnecessary because he's as familiar with the airplane as I am, but it made us both feel better that I'd actually do the same if ever a less-experienced pilot was in front of me.

Once in the air, the most noticeable difference I could tell was that my backside was much more sensitive to turn coordination. Intellectually I knew this would happen and why, but it took being there to really drive home the point that proper turn coordination is a blessed gift to passengers.

Getting the feel for how the airplane handled wasn't that difficult, with one oddity unique to Citabrias. From the front, the trim control is a lever on the left sidewall just under hip high. It falls easily to hand. From the back, while wearing a five-point harness, you cannot reach the trim knob. Instead, it's something you hook with your left toes to trim nose up and step on with your left foot for nose down. Getting subtle adjustments definitely took some work there.

The first landing was ugly. No way around that. We came in to a 38-foot-wide runway with a huge descent rate and bounced halfway to Cleveland. Fortunately nothing was hurt besides my pride.

After that it got better, and by the end of an hour I was beat, but Steve pronounced me fit for backseat operations. I might proudly add that he says I'm one of the few he's ever taught who could get it down in one lesson. That made up for that nasty bounce.

The return home was a breeze, and the landing, a greaser. I wiped her down and put her to bed.

In the ensuing days, I wondered what I would do with this skill. Because I don't instruct, I wouldn't have student pilots in the front. The kids were out, because they were too little to trust with things like starters and mixtures and radio frequencies.

Most of the friends I have are not pilots, and I expected they'd prefer the view out the front better than looking at the gray hairs on the back of my head. In order to do that, though, they'd have to learn the controls I couldn't reach.

Still, it was a fun exercise. It taught me just a little more about flying and a little more about my airplane. And any day I can say that is a good one, indeed.

And in a prophetic sort of way it did pay off. A few years later I found myself the proud owner of a two-seat Pitts, an amateur-built hybrid of S-2A and S-2B. Flown routinely from the back, the Pitts was challenging to learn to land, but I leaned back on that experience in the rear of the Citabria, and found Pitts' landings were not the chore I'd heard they could be.

GREAT WHITE HOPE

My parents were moving furniture after selling their South Florida vacation house, so like a good son I volunteered to help. I planned to hop down to Boca Raton on Saturday morning and fly back that afternoon. If the day got long, I was prepared to stay the night and head home Sunday. My eight-year-old son was eager to go too, saying that he wanted to do something nice for grandpa.

Early Saturday we got up, and I called Flight Service on the way to the airport. A cold front was working its way through South Florida, and conditions from Fort Pierce south were IFR, but it was expected to clear out in the time it would take me to get there.

I pondered driving but decided to fly the Citabria instead because it was beautiful in Orlando and would likely be the same down south by noon. I updated the weather before leaving and found Fort Pierce had cleared out, but the next few airports along the route had not—exactly as forecast. (This was before the days of XM weather in the cockpit, by the way.)

A short while later, I was cruising south along the east coast of Florida at 3,000 feet. My son in the backseat was steering occasionally, fantasizing as only kids can that he was shooting down aliens in some intergalactic war. As I transitioned the airspace at Vero Beach, I missed an omen of what was to come.

A Seneca was coming in on an instrument approach, on a course that intersected mine. The tower's traffic calls to the twin resulted in them calling IMC, despite the fact that their altitude was about the same as mine. I descended to 2,000 feet. I heard the twin's landing but never saw the airplane.

The Vero tower then passed me to Miami Center for flight following, and I transitioned Fort Pierce offshore. Ahead I could see the ceiling was lowering dramatically. I found myself edging down to 1,500 feet, then 1,000 feet, to maintain VFR. The warnings started going off because I was a couple miles out over the ocean and much too low for comfort.

I had descended out of Miami Center's range, so I turned toward shore and dialed in Stuart, the next airport about seven miles south. I asked airport conditions and was informed that the field was IFR. I looked down to consult my chart quickly, and that's when the windows went white.

There I was at 1,000 feet over the ocean in IMC flying partial panel in a sport plane. I made my next mistake. I turned left to make my escape back the way I'd come. If I would have turned right, the shoreline would have helped as it became visible. As it was, I had to complete the turn over featureless gray water.

I broke into the clear after only a minute in the clouds, called up Fort Pierce, and landed for a cup of coffee. I called my dad and told him I was stuck but would probably be there in an hour or so.

Nearly three hours later, I was still on the ground as the front refused to budge. I could have buzzed along the beach at 500 feet and made it legally but decided in this case to exercise a little more discretion. Defeated, I returned home, hoping my dad wouldn't hurt his aging bones by working too hard.

My son was disappointed. So was I. Surprisingly, my dad was not. He was pleased that I'd made the effort but would be around to visit another day.

THE OTHER TEN PERCENT

In the time I've been a pilot, there have been years I've flown a lot and some, not so much. One year, a combination of forces conspired to dramatically limit my cross-country flying and really put the kibosh on my instrument flying. Family responsibilities, ADs on the crankshaft of Lycoming TIO-540 engines, and a few other factors kept me close to home.

However, I'd been doing a bit of testing of both avionics and airplanes, and I was flying the Citabria at least weekly, so my stick and rudder skills were sound. On the gauges, however, was another story.

I enlisted an instructor friend with a 172 to help me flog away some of the cobwebs. My basic flight by instruments was dead-on. Partial panel was no problem, in part because flying the Citabria never involved using a vacuum instrument.

Approach procedures were a little sloppy, but not bad and well within PTS standards. But just when I started getting cocky, the house of cards came tumbling down. I'd just flown an NDB approach to minimums and came out fairly close. We'd briefed a missed approach, and I applied power and started my climbing turn to the holding fix.

At least, I thought I did.

The Skyhawk turned. And I thought the airplane was climbing as best it could. (I hadn't flown a 172 in nearly ten years.)

"Ken, we're descending."

I applied back pressure to remedy.

My friend tapped the VSI, which showed a 200 fpm descent. I scanned the instruments again. There was that descent again. I couldn't believe it. My body felt as if we were in a max performance climb. I started the "trust your gauges" mantra and eventually reconciled the internal feelings with the panel indications.

This wasn't the first time I'd encountered spatial disorientation, but it was by far the worst, and most convincing, episode I'd ever experienced. I finally understood on a visceral and emotional level what must happen in many of the accidents that involve pilots descending to the ground during an approach or missed approach.

As we flew toward the next approach, I kept thinking of what certainly would have happened if this had been for real.

Of course, we were in the air, so it was for real. But fortunately I'd had benign VFR weather and an expert instructor on my side. This time.

This flight served as a wake-up call that hard-won proficiency is a fleeting thing. It was a poignant reminder that instrument flight is particularly demanding. Being able to cope successfully with ninety percent of the task at hand is of no use when it takes that other ten percent to bring the flight to a happy end.

SENSIBLE ALTERNATIVES

I was in the flight planning room at an FBO in Knoxville, on the phone with Flight Service to file an IFR flight plan to Orlando. The weather in the southeast was just lousy with thunderstorms.

The weather radar showed a clear corridor down the center of Georgia if we left *now*. I had briefed the weather on the computer but had called FSS to file the flight plan.

When I told the briefer I wanted to file, he was nothing short of indignant. Hadn't I looked at the weather? Did I know about the hazardous weather? His tone made his real question crystal clear: Do you know how stupid you are for planning to fly a light aircraft on a day like this?

Well, I'm pretty sure I'm not stupid, and I know I don't have a death wish. But the flight certainly looked doable to me. My comrade and I fired up the Mooney 231 and went high, looking to maximize true airspeed, minimize headwinds, and still have a chance of navigating around the big stuff.

As we cruised south, every frequency was stuffed with airliners seeking diversions, but the ones crying the loudest were those west of our position and flying west. Our run down the center was working.

As we passed east of Atlanta, a cell took over Hartsfield. "Everybody on my frequency, I'm sorry but you're screwed," the controller said. "Everybody can expect to hold."

Ahead I could see buildups both left and right of our course, but nothing in front of us. The Stormscope verified the visual picture. As airliners stacked on top of one another at Atlanta's approach gates, we motored happily along and were soon given a frequency change.

Make that almost *everybody can expect to hold,* I thought to myself. So far, so good.

As we neared Orlando, however, a level five thunderstorm stood between us and home. I asked for vectors around the cell. We got into some light turbulence and were in and out of IMC. My companion took the opportunity to snap off the autopilot and hand me the Foggles. I groaned.

"The more you sweat in training," he reminded me, "the less you bleed in war."

OK. It was the end of a thirteen-hour workday. I was on the tail end of seven hours of flying in two legs. I was skirting a very large storm, at night. Sweat indeed.

The flight ended without incident, and I was glad to have had the workout on instruments. But I still think back at the briefer's tone. Was I crazy? I don't think so.

Risk management is part of the game. Although the preflight picture wasn't ideal, it wasn't suicide either. Fortunately that was my call, and not the briefer's.

ROUND, ROUND, ROUND WE GO

One of the keys for operating around airports—or any other time you come into close proximity with other airplanes—is to have some idea of what's happening to you, what other airplanes are doing, and what you're likely to do next.

When a pilot appears to be playing by a different set of rules, it can create a situation that ricochets far past that pilot's immediate action.

I was in an Arrow approaching a Class D airport during a time there was a Cessna 172 making closed traffic and a Bonanza four miles out that had been cleared by the controller to make a straight-in approach. I was coming in on what I was told would be a two-mile base leg. There was also a helicopter warming up on the ramp.

As luck would have it, the Cessna and I approached the downwind-to-base almost simultaneously. The controller was talking to the helicopter as the other airplane and I approached. I was higher, and not yet in the pattern, so I gave way, just like the regulations say.

With the Cessna just in front of me, I started slowing the airplane down. I dropped flaps and gear, but my companion urged me to bring the gear back up because we were planning a test of the emergency extension system. I chopped power and turned out a hair to increase the distance between the Skyhawk and myself, and the controller came on to verify that I could increase separation.

I answered in the affirmative that we'd already doubled our distance. The controller told me to follow the Skyhawk, and he cleared me to land, number 3 behind the Cessna. The Bonanza by now was on short final.

I didn't follow the Cessna because, without warning, the Skyhawk broke into a 45 to 60 degree left bank. Keep in mind we were in a traffic pattern at about 700 feet over a city. The controller saw something was amiss. "Cessna 123, is that you making the turn over downtown?"

The pilot launched into a lengthy explanation of how he'd decided to make the turn because he wasn't sure that I could follow, and right now he was going to go west and extend the downwind.

The controller impatiently said, "All I asked was whether it was you. Yes or no?"

At that point, he cleared me to land No. 1, and the Skyhawk became No. 2. As I turned final, however, it was clear the controller had lost part of his mental picture. He cleared the helicopter for an immediate departure if the pilot could stay clear of the final approach course. The problem was, he used the last three numbers of my call sign when clearing the helicopter.

The helicopter, piloted by a relatively low-time helicopter pilot, did not reply to the clearance. The controller repeated, obviously exasperated, but this time got the number right.

As for the Skyhawk pilot, he may still be circling out there, for all I know.

CONTROL FREAK

"There are a couple of maneuvers you'll never use again," a friend told me.

We had been discussing various maneuvers required on practical tests, and this reaction referred to the chandelle and lazy eights required for the commercial test. My friend, who flew briefly for American Airlines until concluding that flying for an airline wasn't her cup of tea, expressed disdain for the hoops all levels of pilots must jump through for no apparent reason other than to improve the standard of living for flight instructors.

I had to differ on this one. Besides the fact that I think chandelles and lazy eights are fun, they also serve a useful purpose. By learning to handle an airplane at the edges of its performance, the middle ground gets much broader.

In my younger days I was a motorcycle enthusiast who spent more college time riding two-wheelers than I did studying. This conversation reminded me of a maneuver my buddies and I used to practice on our high-performance street bikes. Someone would place a, uh, beverage can on the ground. We'd ride up to it with the aim of scooping up the can, drinking some, and putting the can back on the ground without spilling a drop. Bragging rights went to those who could pick up the can at the greatest speed and return it to its original spot in the shortest time.

Foolish? Maybe. Frivolous? Definitely.

But it did teach very fine control of the machine. Which is exactly the point of the flight maneuvers as well.

You may never need a chandelle to get out of a valley airport, for example, just as you'd never need to scoop up a can while riding a motorcycle. Eights on pylons might be a good way to annoy an ex-wife, but there are better ones. In any event, learning the maneuvers is important to establish your command of the aircraft. Practicing them is just as important.

We all remember things we "used" to be able to do, but can't now—maybe playing a band instrument in high school or chipping a golf ball to within a few feet of the hole. Not using a skill is the surest way to kill it.

So why practice a maneuver that bears no resemblance to the kind of flying you do? The easy answer is that you need to know what you and your airplane are capable of doing, just in case. For example, you might need a high performance climbing turn to avoid terrain and a potential mid-air collision.

For concrete examples, examine the accident record. Consider one pilot landing his iced-up high performance single at 110 knots. He hit hard and damaged the airplane. Fear of stalling led him to keep his approach speed high, but abnormal fear of stalling forced the speed far higher than it should have been.

Practice what you've learned. Unlike the mindless drivel you may have sat through in high school, this stuff you will need later.

THREADING THE NEEDLE

When conducting a weather briefing for a flight north from a small airport in central Florida, the briefer was pessimistic that I would be able to complete the flight as planned.

He cited a convective Sigmet, calling for a band of severe thunderstorms, possibly including tornados, across the entire midsection of the state and extending offshore in both directions. The weather was moving east and was forecast to last much of the day. The briefer's tone strongly implied I should reconsider the flight.

As chance would have it, I had flown south through this band only an hour earlier, aided by Nexrad weather information and a team of controllers who were routing airplanes around buildups like auctioneers. The Strikefinder had gone tango uniform at the start of the trip, and without the Nexrad I likely would have scrubbed the southbound flight. But here I was, about to head north and back into the fray.

The radar imagery showed the storms formed a conga line about eighty miles thick. Vertical development on the biggest cells was well over 40,000 feet. Did I mention the briefer was pessimistic?

I mentioned that I'd just flown the same route an hour earlier and had gotten a good ride with the combination of Nexrad and controller assistance. I thought I heard him breathe a sigh of relief and then we took care of the flight plan.

I did manage to troubleshoot the Strikefinder and discovered the problem was a loose connection at the circuit breaker, and that was addressed in short order. I felt I had the tools to make the flight.

The weather at the departure airport was VFR, but there was a ragged ceiling that ran from about 1300 feet to 1800 feet. I departed westbound, toward Tampa, and the tower controller turned me over to Tampa Approach. For the next ten minutes, I waited impatiently for a break in the controller's rapid-fire instructions so I could check in. Four times I keyed the mike and made the call but got no response. Two times I heard my call sign with a "How do you hear?" but my reply was stepped on. There were a lot of airplanes seeking a lot of diversions.

There was a big gap between storms at my two o'clock, and a turn through it would put me nearly on course. Finally, I got through to Approach, and that controller agreed on a heading that made us both happy.

The ride was not smooth, but far from uncomfortable, and the miles melted away. I was handed off to another controller, who turned me on a direct course for home. I was watching a cell about thirty miles ahead, anticipating it would move out of my way, when the Nexrad went dark.

The Strikefinder showed minimal electrical activity, but it had appeared the cell was building, and visually I could see it towering. Fortunately, the last image I had was that the weather was relatively clear to the west of the cell, and I queried the controller for a deviation. She agreed with my assessment, and away we went.

Soon enough we were through the band and into an area of clear skies with only a broken layer that ran from 2000 feet agl to 6000 agl. From there it was smooth sailing the rest of the way.

In retrospect, I started wondering about my reliance on Nexrad for this flight (and the one preceding it, actually). I had redundancy, of sorts, in the form of a Strikefinder and ATC radar, but the significant shortcomings of each were not lost on me.

The Nexrad unit in question was a Garmin 496, and it went dead for a very predictable reason: the battery ran down, and it could not be powered by ship's power. I knew going down this was an issue, but frankly I thought it would last long enough to make the round trip. It would have too, had it not been for the weather diversions that added time to the flight.

Still, in retrospect I wonder if making this flight constituted an unacceptable risk, given the level of convective activity that struck the area that day. I wonder if the better part of valor might have been delaying the return flight, even if I was not prepared for an RON. In short: Was the Nexrad a crutch I relied too heavily on?

After due consideration, I came to the conclusion that I was both responsible and prudent in my action, given the range of tools at hand. You can bet I'll be addressing the ship's power issue, but other than that I'll log this flight into a web of experiences that seems to grow with just about every trip.

TWO PILOTS, NO CAPTAIN

I snapped into the five-point harness and looked at the back of the front-seater's head. It had been a while since I'd flown an aerobatic taildragger, but the limited view and the on-my-back feel made me long for my old Pitts. Well, this would have to do.

We weren't planning aerobatics on this flight, and so I was sitting on a seat instead of a parachute. The aircraft's owner asked me if I'd like to fly, and I said I did. However, I gave her the front (solo) seat because I *wanted* to recapture that Pitts feeling, at least in some small way.

The plan was that she would start the engine and do the runup, and I would do the takeoff and conduct the flight as PIC. We said we would deal with the landing later, but I suspected I would turn control of the airplane to her when we approached the airport again.

We were flying out of a one-way grass strip. To our left was a bluff that rose about 800 feet above the runway elevation. Ahead was a hill that rose about 300 feet. A river ran along our right. Following lower terrain along the river involved simply jogging slightly right after clearing the 75-foot trees on either side of the runway.

Listening to the common frequency, I heard two airplanes in the pattern and a third inbound over the dam, a few miles north. One airplane was turning base to final, and I saw the second on a close-in midfield downwind. As the second airplane landed, I determined we could get in the air and climb out left, over the bluff, without interfering with the third airplane's arrival.

Runup complete, I took the runway and pushed the stick and the throttle forward. Airspeed alive, gauges green, and tail up, we hopped down the hard grass runway, with each bounce feeling lighter than the last. We reached flying speed, and I dropped the tail slightly. The airplane climbed into the blue summer sky.

Not as well as I'd expected, though. We made it over the bluff, but the climb seemed sluggish even though the airspeed indicator was riveted on Vy of 85. It seemed to be chugging a bit, but I let it ride. The airplane's owner in the front seat had thousands of hours in this bird, and if something was wrong, she'd know.

I chalked up my reaction to two factors. I had a couple hundred hours in a similar airplane, but this one had a less powerful engine. I was also wearing a non-ANR headset for the first time in a decade, so I'd expected the sound to be different.

As we struggled for altitude over the bluff, the radio crackled to life. "You did spins on your last flight, right? Does your airplane always trail smoke after you do spins?"

Trail smoke?

I banked the airplane immediately back toward the

airport, and was somewhat relieved when the front-seater claimed control. We landed without incident. That's when the head scratching began.

The problem was easy enough. The run-up checklist includes a carb heat test, which the airplane passed. However, at the end of the test, the cable connection apparently failed, leaving the carb heat in the ON position and enriching the mixture considerably. The front-seater said she thought she noted a rise at the end of the carb heat test during the runup but admitted there was a possibility that she was disengaged slightly because I was flying.

The larger question is how two pilots, both well experienced in the type of airplane and flying recreationally, failed to detect the problem and continued to climb toward a bluff with the intent of out-climbing it. Remember, a turn to the right would have put us into the face of oncoming traffic.

In every flight, we make certain assumptions about how the flight will transpire. The trick is to keep your mind alert enough to challenge those assumptions at critical times. On this flight, two capable pilots did not challenge their assumptions that this recreational flight would be anything but routine. The mechanical flaw was minor, and the safety of the flight was never seriously in doubt, but the process by which the problem was detected and solved was a bigger issue.

In addition, while I *thought* I was pegged on Vy, in fact 85 was Vy on the Pitts, and not the airplane I was flying. In addition, I was misreading the airspeed indicator four feet away and was really flying at 65 rather than 85, or about Vx for that airplane. The owner read out the airspeed twice, hoping to prod me into lowering the nose, but for whatever reason it didn't register with me, and she took no further action to *make* it register.

It's easy to preach vigilance; it's hard to practice it. In the end, remember the scientists' credo: Expectations cloud observations.

THE BEST-LAID PLANS

We like to think of our airplanes as mini-airliners that will get us from A to B with the dispatch reliability of the airlines, and with the higher cost offset by the greater convenience.

If only.

Most owner-flown aircraft are more at the mercy of Mother Nature than are Part 121 aircraft, especially when flown single-pilot. A recent trip pointed out to me in no uncertain terms that flying by GA aircraft sometimes leaves you wanting more.

My family of five needed to be in the Philadelphia area for a wedding, with my daughter serving as one of the attendants. So she *had* to be there the day before the wedding for the rehearsal. Lacking a firm airline/TSA-imposed time to leave the house, we headed out a couple of hours later than anticipated. The weather between Florida and Pennsylvania looked innocuous enough, but I knew we were cutting it close to get there on time.

Winds aloft were neutral, and as we approached the Washington D.C., area it looked like we'd have about a half hour to get to the wedding site.

If only.

ATC was vectoring us around some active military space and sent us down. As we went into the clouds, we were buffeted by moderate turbulence that lasted almost until touchdown. The turbulence took its toll on my passengers, with my two normally iron-stomached boys suffering motion sickness.

Did I mention the airsickness bags were missing from their normal spots in the seatbacks?

When we got to the FBO, we had to spend some time cleaning the upholstery, and by the time that was done, it was rush hour on a Friday. Long story short, we got to the rehearsal site 45 minutes late. It was truly fortuitous for us that the bride's parents got there fifty minutes late.

The day after the wedding, we had planned a side trip to the Maryland shore, but a massive low-pressure system had parked there, and we decided to stay put for a day or two. I got the opportunity to provide a friend's twenty-something-year-old son with his first ride in a GA airplane, and the look on his face really refreshed my enthusiasm for why we do this flying thing.

Too soon, it was time to head home. I'd been keeping an eye on Tropical Storm Edouard, which came ashore in Texas earlier in the week. I knew the low-pressure system would work its way east, and, tropical storm or not, it was not something I was eager to tangle with.

Unfortunately, that low-pressure system and the low over Maryland merged into a stationary front that battered the southeast for several days, extending pretty much entirely the length of my return window. We delayed our return as long as we could, but we decided to set out on Wednesday to return home.

I figured I could head for Charlotte and then run west of the front until south Georgia and then pick my way through the embedded thunderstorms with a combination of XM weather, Strikefinder, and eyeballs. However, when the briefer mentioned tornado watches throughout north Florida until well after my anticipated arrival time, it dawned on me it was time to think of some alternatives.

We could press on, with the near-certainty that we'd have to land short and wait a few hours until things sorted themselves out a bit. By now, however, my wife had to get back for important board of directors' business at her large company.

In the end, we headed west to Charlotte, but instead of turning south for home, we continued west to my mother-in-law's farm in Tennessee, where we left the airplane, borrowed a car, and drove home.

During the late-night drive, the weather was better than forecast, and I began to second-guess my weather wariness. However, the next day I found the storms had killed the motor in our hangar door, and if we'd have gotten back in the wee hours in the airplane, we'd have been stranded there, unable to raise the door to get the car out of the hangar.

So in the end it turned out okay. Sure, I had to return to Tennessee to get the airplane, but we'd been planning to get the car anyway, so a trip was in order regardless. Sometimes it just shakes out that way.

GA isn't perfect, but we got home, more or less on time, with a minimum of inconvenience and hassle. And anyone who's been stuck on an airliner overnight in a remote city or on the ramp in a snowstorm has been in just as bad a situation, maybe even worse.

PART II
THE SANITY OF FLIGHT

ARE YOU NUTS?

Since the dawn of flight, pilots have been forced to validate their infatuation with aviation by giving a convincing answer to the simple question, "Are airplanes safe?"

It's a loaded question that addresses a complex set of dynamics, and like most simple questions, has no simple answer. No matter what the pilot answers, the response is suspect.

If you answer yes, then you are deemed either a liar whose aim is to generate undeserved approval for an avocation that's clearly dangerous, or a reckless daredevil with either a tiny brain or a death wish. After all, the evening news is routinely adorned with tales of airplanes—and their occupants—lost to disaster.

If the answer is no, you are simply an idiot for pursuing such folly.

Give the proper answer of "it depends," and you are a fence-sitting simpleton trying to figure out a way to bastardize statistics to support whatever far-fetched claim you're about to utter.

So try this. Put yourself into the shoes of the general population and answer this: "Is air force flying safe?"

(Military pilots, you're excused from this exercise because you already know the answer.)

Most pilots recognize that military flying is decidedly different from general aviation flying, even during ordinary times. The high-performance aircraft are routinely operated near the edges of their envelopes. Maintenance is intense, but the pilot training and proficiency also exists at levels that put general aviation pilots to shame.

It shows in the statistics. General aviation trainers had an accident rate of about 4.2 per 100,000 flight hours during a recent five-year period. Four-seat, fixed-gear cruisers recorded 3.9, and medium twins had about 3.

Compare that with the F-16, which suffered 3.70 Class A accidents per 100,000 flight hours during those five years. The A-10 recorded 1.68, and the F-15, 1.94. The F-117 was the highest of the lot, suffering 6.14.

These figures are for Class A accidents, which are those that involve fatalities, loss of the aircraft, or $1 million in damage. Roughly 40 percent of the air force's Class A accidents are engine related, and nearly half are operational.

The safety record is even more noteworthy among trainers. Consider that the air force's venerable T-38 trainer had only one pilot fatality in an entire decade. During that time, the aircraft flew more than 1.6 million hours. The T-37 flew nearly 1.8 million hours in that time and lost two pilots.

The upshot of this analysis is that general aviation flying is safe. Or it's not. To reduce the world of flying to a simple yes or no is to reduce art to a box of crayons or cuisine to a bag of carrots.

Is flying safe? That's easy. It depends.

THE DEMON IN CHAINS

Buried somewhere inside every pilot, a demon impatiently waits to be set free.

Many keys will unlock its chains. Complacency, carelessness, arrogance, or foolishness will do quite nicely, thank you. Once free, the demon can wreak havoc on our airplanes, our lives, our families, or total strangers.

Proficiency and attention to detail reflect the careful pilot's task of keeping the demon chained. Because we're human, we can't possibly fly each flight perfectly. We might bust an altitude, blow a clearance, forget a frequency, or mistakenly taxi onto an active runway at an unfamiliar airport. It's our job, however, to make sure the little mistakes don't snowball into big ones. That's when metal gets bent and bones break.

What we put into flying often dictates what we get out of it. A professional attitude leads to professional results, while a cavalier approach makes the outcome much less certain. General aviation pilots are unique among the countless groups of enthusiasts in that even the most pedestrian participant puts extraordinary care into learning from the mistakes of others. We can learn through the experienced eyes of those who have been there before. We can learn through practice and analysis and introspection. The important thing is the desire to fly better, safer, and with more confidence on every flight.

Rare is the pilot who does not endeavor to develop ways to complete each flight safely and without incident, strenuously trying to find ever more tools to keep the demon chained within. Most aviation publications carry instructional articles, and those articles are frequently pinned to accidents contained within the comprehensive archives of the National Transportation Safety Board. Try finding that in a magazine on boating, skiing, or even guns.

Federal regulation of aviation is considered to be "written in blood," because the aftermath of every single accident includes an analysis of how to prevent it from happening again.

But the demons are many, both known and unknown, and each seems eager to find blood of its own.

GREEK THEATER OF FLIGHT

In ancient Greek theater, playwrights used a mechanism called *deus ex machine*—god from the machine—to lower an actor to the stage like a god descending from the heavens to right whatever wrong was happening onstage and return order to the universe.

Divine intervention may have worked on theatergoers thousands of years ago in Greece, but it doesn't do diddly in aviation. Unfortunately, too many pilots rely on blind luck to save their skin.

Spend a little while reading the NTSB accident reports, available on the agency's website at www.ntsb.gov, and you'll find that boneheaded pilots are not in short supply. Their ill-conceived maneuvers generally fall into predictable patterns, with the occasional creative individual who finds some new way to harm someone with an airplane. The worst part is that they take innocent people with them.

Top on my list of loathing are pilots who try to impress someone with aerial maneuvers, usually aerobatics, when they don't have a clue what they're doing. The pilot can take his poor judgment to the grave if he wants, but why take an unwitting passenger along for the ride?

Buzzing and low-altitude aerobatics are reckless driving. Experienced pilots and very lucky rookies can live through it. Others do not. There are many cases in which middle-aged student pilots decide to take a spouse or child for a ride. Once airborne they get the urge to show off like an adolescent in a sports car.

Also high on my list of annoyances are pilots who forget that airplanes are driven by internal combustion. Time and again we hear of people who fly five-hour legs and declare an emergency when the engine quits two miles from the destination. If time is so precious that you can't afford a half-hour fuel stop, why are you flying a light plane?

Next are VFR pilots who take off into bad weather. I can make some allowance for those who encounter unforecast bad weather, but those who knowingly charge into the abyss deserve what they get.

Of course, accidents do happen. But for a little luck there are a few times when I might have found my own actions regurgitated in an NTSB report. But there's a fundamental difference between pilots who have accidents because of true misfortune and those who are simply irresponsible.

What does their unconscionable behavior cost the aviation community? Plenty. A large percentage of small plane crashes make the evening news or the morning paper. Non-fliers who see the reports cast a suspicious eye at small planes and doubt the sanity of those who fly them. The perception of danger is out of sync with the actual risk, which, although relatively high by today's standards, can be acceptably managed.

As a community, aviators need to spread the word that safe flying is like safe sex. Ignore it once, and it may be too late for some Greek god to come to your rescue.

THE FOUR DIMENSIONS

Faster, slower. Higher, lower. Bigger, smaller. Expensive, cheaper. Like the four dimensions physicists use to describe the universe, these four dimensions describe the universe of general aviation airplanes.

The speed, altitude, and size of the airplane you fly determine the kinds and levels of risk to which you expose yourself. The most extreme example is one of a military pilot skimming the treetops at attack speed, who is in a decidedly riskier spot than that same pilot ferrying that same aircraft from one base to another in the flight levels.

The same could be said for general aviation airplanes, though perhaps the extremes are closer together. Flying a J-3 Cub at 1,500 feet agl and 65 knots is far different from flying a Malibu Meridian at 29,000 feet and 230 knots. Weather patterns are different, stresses on the airplane are different, or the pilot's proficiency is taxed in different ways.

You may be comfortable dealing with ice, but how about your forward slips into a short, soft field? What about flight planning? Do your avionics all work? Checked the hoses around your turbocharger lately?

The point is this: A pilot may put a lot of emphasis on instrument training at medium altitudes because he or she perceives the greatest risk to come during long cross-country flights. Many of those skills, however, are negated when the same pilot opts to spend a morning flying a vintage taildragger out of a grass strip. The converse, of course, is also true.

But there is also a wildcard, and that is money. As any aircraft owner knows, buying an airplane is just entry to the club. Maintenance and continuing training are also part of the mix. Skimp on any one of the three, and you're asking for trouble.

Spring is the time of year when many pilots get into trouble. Skills atrophy over the winter. Pilots who limited their flying to local flights on nice winter days now are faced with blustery spring winds and the arrival of thunderstorm season. Those who blasted cross-country in de-iced twins may be itching for some nice-weather, low-and-slow flying.

Airplanes that have been tied down outside over the winter may have hidden damage from animals, water, ice, or wind. Many of these airplanes will be for sale as the owners try to capture the aerial lust of others or sate their own. Those that were annualed last spring may need more than a good preflight before lighting the fires.

The universe of airplanes, just like the skills of pilots or the checkbooks of owners, is subject to limitations. Knowing yours in all dimensions is crucial to keeping the metal away from the pavement.

SHOTS IN THE DARK

Like most pilots, I wish I had a nickel for each time I've been asked the question, "Is flying safe?" And it's a little irritating to have only a couple of responses to offer.

Option 1: "It depends." Yeah, that sounds wishy-washy. Put yourself in your passenger's brain for a second. "If this guy can't answer a simple question like that, why in the world should I think he's got the smarts to pilot this airplane?"

Option 2: "Of course it is. It's my butt up there too, you know, and I wouldn't do it if it wasn't safe." Okay, now the pilot sounds delusional. There are "always" newspaper and television reports of airplanes crashing into houses, fields, and mountains. There's fire, death, and destruction. Safe, my eye!

Option 3: "About as safe as riding a motorcycle." Oh great, the pilot is either one of those kamikaze urban bikers who do wheelies away from stoplights, or else he's a Hell's Angels wannabe. Pass on this flight, thank you.

Clearly none of these are acceptable to pilots who want to justify both their passion for flying and their common sense for accepting the risk involved. But the simple matter is that there's no answer to that question.

But there could be.

The biggest stumbling block in getting a real-world handle on the safety of general aviation flying is the FAA. Yes, the bureaucrats who seem to want to micromanage the lives of pilots almost down to "boxers or briefs" have neglected one little item. In the regulators' quest to ascertain that airplanes are uniformly safe, they have neglected to provide the ammunition to answer the very question they presume to ask.

This is a little like asking Brett Favre to throw pass after pass, yet never allowing him to see if the last one was complete.

Look at any of the statistics presuming to assess the safety of airplanes, and you will (or should) see a disclaimer that you're getting ballpark figures because no one has yet figured out how to put a good number on the exposure pilots have to the various risks. So we guess.

The FAA distributes surveys to a sample of pilots occasionally that ask how much the pilots fly. Those responses are extrapolated to the pilot population as a whole. The trouble is, people don't generally know, so they guess. And they usually guess high.

Ask your AME about the number of hours reported on medical applications. Some guys report one hundred per year five biennia in a row, but their logbooks show six hundred hours. Ask your insurance broker about the number of people who have reported seventy-five hours in the last year, every year since they bought their airplane five years ago, but they have only two hundred hours time in type. Ask your mechanic how many owner-flown airplanes that come in for annual inspections haven't racked up enough hours in the last year to need an oil change.

To have the data necessary to get a handle on general aviation safety, the FAA would need to start keeping track of airplane usage. They could have mechanics report the number of hours flown in the last twelve months after every annual inspection, tabulate the results by model and type, and make it freely available to the public. Maybe then we can start to get answers that make sense.

I'm really the last person who wants to see more regulation, and I can't think of anyone outside of the federal government who would like to see more FAA bureaucracy, but the bureaucracy is driven by the perception of safety, and right now that perception is less certain than an al-Qaeda confession.

PROTECTIVE CUSTODY

It's a gross overgeneralization, but pilots seem to have a morbid curiosity about what brings airplanes down. Many people read an accident report and come away with the conviction they'll "never be that dumb" or conclude, "That guy sure had a run of bad luck."

It's easy to look at operating an airplane, risks and all, from the perspective of how it affects *me*. However, like most things in life, there is more to it than that.

I once received a letter that all pilots should read. Not because it's a situation that any individual is likely to find him or herself in, but because anyone with a pulse would stop and think about the aftermath of an accident.

A reluctant spouse took pen in hand and wrote, in part:

"We lost a dear friend in a small plane accident, and I have watched that person's family struggle in nightmarish ways over his death. Like our friend who died, my husband is also a successful businessperson, one who began taking risks early in life. Although his risk-taking has had some financial benefits, other decisions have brought pain to us. I believe to a great degree that he feels invulnerable, and that makes me afraid."

For all the spouses who worry every time the flight bag appears, here are some strategies for making sure that hardheaded pilot of yours lets only the airplane's wheels touch the ground.

There are several things a reluctant spouse can do other than sit home and worry. First, take some flying lessons or at least a pinch-hitter course yourself. That will give you more insight into what constitutes dangerous behavior and will leave you better equipped to challenge your pilot's flying decisions and exposure to risk. If he's determined to fly despite your feelings, encourage him to get an instrument rating. VFR flight into bad weather is a leading cause of crashes.

If he uses the plane on business, make it clear that it's okay to spend the night in a hotel if the weather closes in or the trip home might be dangerous because of fatigue, mechanical problems, or illness.

If he rents an airplane, make sure he rents from the most reputable place around, not a bargain-basement outfit. If he owns a plane, insist on top-notch maintenance. Some owners dump all their maintenance money into slick paint and fancy interiors and skimp on the parts that keep them up in the air.

Insist that he stay proficient. It's more likely he'll be in harm's way if he's an occasional pilot. Get him to agree to go up with an instructor at least once every six months.

Make sure he has a life insurance policy that does not contain a general aviation exclusion.

Finally, if his risk-taking is extreme, and he is, in fact, a reckless pilot, turn him in. Losing his certificate for a few months might give him the incentive he needs to reconsider his decision-making strategies.

And pilots, if your better half comes to you with these suggestions, think about who you're really protecting and govern yourself accordingly.

NO GEAR

Aviation is the breeding ground of many an old saw. Some have grains of truth, others may more accurately be called wishful thinking inspired by jealousy.

"There are two kinds of Mooney: one with fuel tanks that leak, and the other with tanks that will."

"The second engine on a twin is just there to get you to the crash site."

"The greatest wastes are runway behind you, altitude above you, and fuel back home in the fuel truck."

Of course, such sayings are spawned by the grain of truth they all contain. I have to admit, I scoffed when I was going through my complex training and heard the one about two kinds of retracts: those that have landed gear up and those that will. Surely that had to overstate the likelihood of a gear-up landing, I thought. They wouldn't make them if they were that unreliable.

For several years I flew with the most serious landing gear-related "incident" an occasion when an insect-clogged pitot tube prevented the gear from retracting. The landing gear was such a part of speed control on my Mooney that I wondered how anyone could get slow enough to land gear-up.

Fast forward several years. The Mooney was gone, victim of a growing family, and I was still shopping for the next steed. I flew every plane I could beg, borrow, or rent that had more interior room, looking for just the right model.

All of that variety did a lot for my ability to think aerodynamics but put a serious crimp in those rote operations that come so quickly in the pattern.

One bumpy day I was flying a Saratoga that, as it turned out, was for sale. I was accompanied by a retired USAF pilot and now a CFI. From the start, this was clearly an instructional flight and not a sales demo, so we'd addressed the cockpit resource management issue that brings some flights to grief.

Between the bumps, busy traffic pattern, and the instructor's desire to efficiently impart knowledge about the airplane and its systems, our GUMP somehow got bumped.

Base became final, flaps were down, the descent rate was prodigious to my Mooney-trained eye. I had just started to round out for what was sure to be a squeaker when some blessed soul in the tower put in his two cents worth.

"Saratoga 123, no gear."

We got those 300 horses galloping, pitched up, cleaned up, and went around, having dodged the sound of prop on concrete by about fifteen feet.

Lessons in humility are hard to swallow, but this wasn't nearly as bitter as it could have been. We went around the pattern in silence. No more Q&A about systems or emergency procedures or what ifs. We landed uneventfully, with each of us having checked the gear at least six times apiece, and called it a day.

THE MANY FACES

We all know of someone who started out as an aviation enthusiast but fell by the wayside somewhere along the way.

For some, the cost of flying became either unaffordable or too much to justify to a spouse or oneself. For others, the time it takes to remain a safe pilot was too much to ask. For others, the novelty simply wore off.

One person I know traded up from a Bonanza to a Baron shortly after I met her. She was very proud of her "new" bird and liked to boast of its performance, especially when compared to the Mooney M20J I owned at the time.

Over time, however, her boasting lost its enthusiasm and eventually turned into sour references about the cost of maintenance and engine overhauls. Although I was tempted, I resisted the urge to parry with a comparison of my 201's cost per knot.

The final indication of her exasperation, however, came when she rented a car to make a time-critical 450 nm trip rather than fly her airplane. Predictably, the airplane was soon sold.

Another friend was a 400-hour instrument pilot when he decided to get a multi-engine rating and began using his skills extensively in his business, aided by the fact that his employer owned a Cessna 340.

Before long, however, the business relationship soured, and my friend launched his own company. Cash was short, but time was even shorter. He grounded himself about two years ago. The good news is that he is cautiously talking of buying a single and putting it to work with his company.

The most disappointing example of giving up involves a man who was a student pilot at the same time I was. When we met, I had about twenty hours in my logbook, and he had about forty-five. A year later, I had about one hundred hours and he had about ninety.

The catch was that he was still flying on his student ticket because he had never bothered to schedule a check ride. Every month or so he would rent an airplane and fly around for a while—sometimes on cross countries, sometimes on extended "local" flights.

When he finally stopped pretending to try to get a license, he had more than 120 hours.

These examples show that the dynamics that either bind us to or eject us from the world of aviation are complicated and ever changing. The key is that each pilot should be able to look in the mirror and say, "Where am I going?"

The rules are simple. If you're going to fly along, do it right. Fly smart. Stay proficient. Understand what you don't know.

Aviators are generally a congenial bunch and generally acknowledge the pressures we all face to stay on top of both our flying and our finances. When someone clips their own wings, we all understand that, but for the barest of fortunes, there we go too.

FLYING AT WORK AND PLAY

In flying, like in any endeavor, you get out of it what you put into it. Some people consider flying a hobby, and they're perfectly happy tooling around in a Cub or a Husky or a Skyhawk. To them, the world up there is one of peaceful vistas and three-dimensional freedom.

Some people find flying defines their lifestyle. They may make their living at it or use general aviation as a serious transportation tool, flying in all sorts of weather, sticking to a schedule and dealing with minor adversity before it becomes major.

Still others are consumed by the passion of flying. To them, flying is a challenge to be met, a task to be mastered and an adventure.

Most people, of course, share all of these traits to some degree. Their task is to balance the challenge with the lifestyle and still have some fun.

Once upon a time, I determined it would be possible, maybe even desirable, to make a foray into multi-plane ownership. The Piper Lance in the hangar was big enough and comfortable enough for those trips to grandma's house. It was reasonably fast, thanks to a few LoPresti speed mods. And it was almost flawlessly reliable.

It was also achingly dull to fly.

The fix for that, of course, was to get another plane for fun. Being based in Central Florida, the prospect of a seaplane had definite appeal. Chasing airboats, going fishing, exploring the hundreds of lakes that dot the local landscape would be a lot of fun. And there's no substitute for the looks you get when you taxi an airplane up to a dock at a marina to grab a bite for lunch.

On the other hand, who can go to an airshow without fantasizing about flying a Pitts or an Extra or even a Stearman home? For me, getting into the competition wasn't the appeal. Winding around the sky in anything other than the Lance's straight and level looked just fine.

Call it gentleman's aerobatics, which compares to airshow flying roughly the same way a pickup game of touch football compares with the NFL.

Regardless of which way the "fun flying" decision was to go, I recognized that the important aspect will be to make sure the three legs of the flying stool are all in balance.

Part of it is a hobby. Part is a lifestyle. Part most certainly is a passion. Keeping them in balance is the way to keep your head in the flying game, to stay alert for the unexpected, and avoid the complacency that invariably sets in when we become too set in our ways.

Try it yourself. If you're a hobbyist flier, take a trip. If you're not instrument rated or don't feel comfortable, bring along an instructor. Some of them won't even expect you to buy their lunch (although you should). If flying is your lifestyle, rekindle your passion. And if airplanes are your passion, make sure it's tempered by the realities that can otherwise hit too close to home.

MEET GEORGE

The quest for enjoyment while flying depends on having the right plane for the right mission. For some, a J-3 Cub and a grass strip provide the ultimate aviation experience. Others find satisfaction in a Baron at 18,000 feet, or an F-16 going straight up.

Every spring and summer, the air show season awakens the lust in people for hardware they'd like to see in their hangar.

The problem with airplanes, of course, is that design compromises must be made that force airplanes into niches. No single airplane can do everything, or even close to it, and so all aircraft designs are, from time to time, bound to be found wanting.

A seaplane is too slow to take on those long business or vacation trips. A sporty plane probably handles too poorly in IFR conditions. A cross-country plane may be uninspiring or too expensive to use for fun.

Nimble or practical? Flashy or utilitarian? Cheap or sophisticated? The list goes on and on.

For most people, there is no perfect solution. That is,

unless you're one of the lucky few with the cash flow to support a harem of airplanes.

Meet George. He went to Oshkosh to look for a Cub to bring home as his new toy. Look in George's hangar. He already has a Lake amphibian and a high-performance single. There are some who would look at a personal fleet of airplanes as wretched excess. There are others who would consider it nothing more than a good start. (A few airplane-smitten celebrities come to mind.)

For most of us, there's only the wistful thought that maybe someday we'll be lucky enough to have the keys to three airplanes dangling from the key chain in the car as we drive to the airport. There was a time when I was making that cautious assessment myself, wanting to add an aerobatic airplane to my fleet of one. I crunched the numbers and decided we'd be able to do it if 1) I could convince the kids to go to public schools and in-state colleges, and 2) the dinner menu was heavy on macaroni and cheese for at least seven years.

When it came down to it, however, my better half put her foot down. "You'll have a second airplane only after I have a beach house" were her exact words. That second airplane remained a fantasy. Nevertheless, I do occasionally mill about the aircraft displays at an air show or peruse one of the aircraft-for-sale websites, thinking about some sweet, sexy taildragger. Maybe I can get someone to lend me their Extra for a couple of weeks.

TWO DOWN, AMONG THE MANY

July 1999 attained a bit of infamy in the aviation world. Not only did New York (and the rest of us) lose JFK Jr. to a light plane accident, National Air and Space Museum chief Donald Engen died when the motorized glider in which he was riding suffered an in-flight breakup.

During those two events, popular media managed to present startlingly inaccurate information to the general public about the safety of light plane flying, especially in its round-the-clock coverage of the search for Kennedy and his passengers. The misconceptions born out of that coverage stayed with us long after the average American forgot what VFR means. Many people stubbornly concluded what they'd always suspected: that light planes will fall from the sky at the slightest provocation.

In the weeks following the Kennedy accident, I tried to do my part, appearing on camera for MSNBC and Fox News, granting interviews to *The Wall Street Journal,* writing letters to *USA Today,* and appearing as a guest on several public radio talk shows.

Aviation accidents predate the Wright brothers, and

it's unreasonable to assume that, even after a hundred years of practice, mankind will figure out a way to stop them entirely.

The best we can do is equip our airplanes with the best safety gear we can think of (and pay for), keep our skills as finely honed as possible, and practice the best judgment we can. Then just let the chips fall where they may.

Interestingly, it was that same month that Cirrus Design began deliveries of its SR20, known to the rest of the world as the airplane with the parachute. Cirrus went on to become the highest profile manufacturer of piston singles, dethroning even the legendary Cessna Aircraft as the industry icon.

Cirrus represented new technology the same way Cessna embodied the traditional, despite the fact that many of the components were tried-and-true. Cirrus would go on to champion a number of safety improvements, but it was the SR20 that launched the company and proved that pilots really would respond to a new aircraft if it was compelling enough.

The juxtaposition of Kennedy and Cirrus clearly showed that, while general aviation would continue its gradual improvements in safety, it's up to pilots to ensure that one of the oldest tricks in the book—crashes—doesn't turn into a new way to feed the media machine.

TWO TIMES TWO

A pilot friend of mine is one of the most knowledgeable people I know when it comes to IFR flight and regulations. He has the FARs cold and knows the information in the AIM better than some people know their birth certificate. He flies VFR whenever he can.

Another pilot I know has to look things up occasionally. He admits to a certain befuddlement about some of the rules, and deep down I suspect he's a little leery of shooting approaches when the weather is at minimums. He files an IFR flight plan in the clearest weather and accepts non-direct routings as the price he must pay to be solidly in the system.

Both guys are skilled pilots. Both have solid stick and rudder skills, use the radios efficiently, and grease far more landings than they bounce. The main difference between them is attitude.

Mr. VFR pilot has an independent streak a mile wide. He follows the rules to the letter, unless it suits him not to and there's no chance of a violation. Then he'll bend them to suit his purpose. He thinks he's been around enough that he knows how far he can bend them without getting into trouble.

Mr. IFR pilot is a team player. He's willing to pay a

small price if it helps everything move more smoothly. He follows the rules to the letter, unless he forgets one or something falls through the cracks. He thinks he's being conscientious, but in some ways he uses the system as a crutch to make up for his perceived shortcomings.

A good instructor can teach just about anyone the mechanics of flying an airplane. While some have a natural feel, and others have to acquire it, anyone of normal intelligence can learn to take off, land, navigate, and recover from stalls. As with my acquaintances, attitude is the great divider.

Some people demand perfection. They roll smoothly, and power changes are nearly imperceptible. Others enjoy the crispness of high performance maneuvers and fly the airplane to the limits of what its designers intended it to do.

The question becomes, who would you rather fly with? The answer, of course, should be neither. Or both. For each approach has its strengths and weaknesses, and only by knowing the context of the situation could we show one to be a better choice than the other. Ideally, one could take the best of each pilot's attitude and merge them into one highly capable pilot.

Eventually I put these two guys in a cockpit together. The actual flight, or rather flights, as it involved two legs, is a story best told over a few cold ones, but for now we'll just leave it at the fact that the whole experiment ended up going nowhere pretty fast.

BE IT RESOLVED ...

It was the end of December, and unlike most years, this time I actually planned to make some resolutions, including how I'd approach flying, the airplane, and my peers in the coming year. Maybe you will too, even if you're reading this in July.

Resolutions, of course, are only worthwhile if you plan on keeping them, so I think it's valuable to consider the kind of flying you do and the ways you think you can improve. Any pilot, from student to veteran ATP, should be able to come up with a few. If not, they're not nearly as good as they think they are.

For my part, here are an even dozen flying resolutions I came up with. I have a few others, but I don't think they belong in print just now.

- I resolve to give my mechanic more leeway to fix things that might break before I'm ready for them to.
- I will make my weather briefings more comprehensive and take better notes, even if I think I already know what the weather is like.

- I will learn from the mistakes of others instead of just shaking my head and telling myself I'd never be that stupid.

- I promise to worry less about what might go wrong, but I'll make sure I know the full capabilities of my airplane just in case.

- I am committed to fly at night more regularly during the summer, so I won't feel as rusty when fall comes.

- I will treat the CFI who gives me my flight reviews with the respect a professional deserves. (Which I do already, of course.)

- I resolve to check out in at least one new type of aircraft this year.

- I promise to treat controllers with the respect due a professional, even during those times when it seems they're not according me the same courtesy.

- I plan to introduce at least one landlubber to the joys of flying this year.

- I resolve to more thoroughly understand operation of the avionics installed in my airplane, the relationship among the boxes, and the various failure modes.

- I resolve to stick to my personal minimums for instrument approaches more carefully.

- I promise to give my passengers the smoothest ride I can, unless I am flying something aerobatic and they beg me to scare the dickens out of them. Then all bets are off.

Now it may seem that some of these resolutions will kill the joy of flying, but in fact I think it will awaken even more of a passion. Nothing kills enjoyment like uncertainty, and with these resolutions I hope to address my weaknesses, learn a little bit more about flying, and spread the wealth a little.

I just hope I do a better job keeping these than politicians do at keeping their campaign promises.

Make up your own list of aviation resolutions. Promise yourself you'll learn something about yourself and your airplane that will make your flying safer and more fun. Good luck.

MIRROR, MIRROR ON THE WALL

There's a pilot I know who is a perfectionist to such an extreme it makes me sick.

You'll never catch the guy with outdated charts. One time the database on his IFR-approved GPS expired by a couple of days and he wondered if he should fly his entire route by VOR.

He has an IPC every six months, regardless of how much actual instrument time or how many approaches he's had since the last one. Even though he holds a private license, he attempts to fly to ATP standards at all times. His instructor says he usually succeeds. When he doesn't, he's depressed for a while.

When it's time for a flight review, he challenges the instructor to teach him something. Often he ends up teaching the instructor a thing or two as well.

Ask him about the effects of altitude or weight on maneuvering and stall speeds, and he'll spit out the numbers for his airplane without a pause.

He insists on a well-maintained airplane, even though he grumbles about the cost sometimes. I once caught him flying with a loose screw on the mounting bracket of his auxiliary microphone for several months, but it was fixed at annual.

His preflights are meticulous, although he might tell you that he really ought to stick his head farther up in the gear to check the linkages just a bit better. Some would call his weather minimums too conservative. He says there aren't many destinations he thinks are worth dying for.

If he hasn't flown in a couple of weeks, he says he gets butterflies that would impress even John Lennon (who was known to throw up before going on stage). You'd never know it by looking at his technique. People who fly with him often remark that he flies like someone with ten times the hours.

He says he's never intentionally busted an approach minimum, and for some reason you believe it. Then he says he's only flown a handful of missed approaches, and somehow you believe that too. When he finally says he's only scrubbed a few flights on account of weather in about fifteen years of flying, you wonder what kind of airplane he's been flying and where he's been flying it.

His traffic patterns are crisp, tight enough to be safe, and wide enough to accommodate traffic and workload.

I think it's the goal of most pilots to see themselves in at least a few of these statements. It's a given that if you see yourself in all of these statements, you, too, would make me sick. If you see yourself in none of these statements, or if you read them and say, "What kind of a wuss is he talking about?" you might want to reconsider your choice of hobbies.

This describes a real person, and one who makes flying as much a part of his life as any non-professional, nonretired pilot can. He uses his airplane for serious personal and business travel. He borrows or rents sporty airplanes when he can, just for the heck of it.

And sometimes, when I get really thoughtful about flying, aging and my mortality, I think of my wife and kids and say to myself, "I want to be more like him."

COVETING SOME SUPPORT

I could feel the comforting vibration of the two 350-hp engines in the soles of my feet. Through the acrylic in front of me, the visibility was poor as we moved in and out of light rain. The moving map on the GPS showed we were on course, with an ETA of about ninety minutes.

I settled back into the seat, relaxing a little despite the light chop. The rest of the family was behind me, snacking or something. I didn't bother to turn around and look. With the autopilot engaged, I took some time to extend my panel scan. The fuel tanks were reading a little less than half full, and I'd found the gauges to work reasonably well in the two years I'd owned this baby.

I played around with the radar a little, adjusting tilt and range. There was plenty of rain ahead and to our west but no thunderstorms.

My wife came forward and sat beside me, staring at the murk. We knew the clouds went right down to the deck, and she seemed slightly nervous because of the poor visibility. But we were running strong and there was nothing to hit, and she began to relax.

The radio was eerily quiet, and I was already thinking ahead to our destination. I'd wanted to get in before dark, but as usual we'd gotten a later start than we'd anticipated, and now it was clear we'd be arriving after sunset. What bothered me the most was that we had to make our approach through some challenging geography in lousy weather.

I reached over and bumped the throttles up a hair, as if the three minutes it would shave off our cruise time would be enough to matter. My mind went back to that approach. I could see the pass we needed to get through on the paper map in my lap and on the GPS. But I was still nervous. This time, there would be no friendly voice on the radio offering vectors, nor did I have an approach plate.

The radar showed a small blip in our path. I banked slightly to avoid it. Looking out the window as we passed, I saw a crab pot pass to port at about 19 knots.

I was piloting our 38-foot motor yacht, *First Draft*, down the west coast of Florida from St. Petersburg to Captiva Island, where we'd be staying during the kids' spring break. In many ways, this trip was similar to those we'd taken so many times before by air. It always takes longer than you think to get underway, and you're always wishing for a few knots more cruise speed, even though it doesn't make much difference to actual travel time.

This trip made me appreciate flying all the more, despite the fact that the boat was much roomier and had all the comforts of home. I found myself aching for a nice, simple instrument approach that would take me to a paved and lighted runway, instead of staring at the prospect of making a night run through an unfamiliar pass mined with shifting shoals. The GPS was nice, but radio navigation aids like an ILS or vectors to final would have been great, since I wasn't 100 percent sure where the marina was once I passed from the Gulf of Mexico into the Intracoastal Waterway.

It's human nature to get fed up with "hold for IFR release" or lousy vectors that take you on a cross-state tour to line you up with a downwind runway. We like to gripe about waiting for the fuel truck and the steady parade of chart updates that clog our lives. However, pilots have many resources at their disposal that would make other enthusiasts, such as boaters, green with envy. Even the worst electronic navaid is better than trying to find a channel marker in the dark, rough waters with a handheld spotlight. Well, maybe an NDB hold with a strong crosswind is close.

The facilities developed for pilots, such as navaids, ATC, charts, and Flight Service, are there to enhance the capability of the airplane along with your safety in operating in airspace that might suddenly become crowded or choked with thunderstorms or ice. While those assets do carry with them some responsibilities in terms of maintaining proficiency and showing some respect for the rules, think how hard it would be to search for a water tower painted with an airport designator with only a landing light.

SANITY LOST

After nearly fifteen years of flying and owning several traveling airplanes, I took leave of my senses and bought a factory-new, VFR sport plane. I knew life was going to be different.

For years I had filed an IFR flight plan for virtually all flights out of the local area, even on trips of 150 miles in clear weather. I was very comfortable in the ATC system and had learned how to work it as well as how to work in it.

Having the Citabria was a blast. I used it for some light cross-country duties during fair weather and relied on borrowed or rented airplanes when the clouds were low. Then the FAA went and changed all that.

A serious AD on Lycoming TIO-540 crankshafts grounded a large number of airplanes, and the limited parts supply kept them grounded for a while. Suddenly it was not as easy to find a set of wings for IFR flight. I cautiously investigated adding IFR capability to my Citabria.

Stop laughing. It's not a ludicrous idea, as long as you don't expect a fully equipped Citabria to be a long-haul machine. But by the time I got done with my analysis, I decided maybe the whole rental thing wasn't so bad after all.

To equip the Citabria for minimal IFR capabilities

would have involved adding a heated pitot, clock, nav/comm, CDI, vacuum system, attitude indicator, directional gyro, and antenna. For the $7,000 or so, I'd get the ability to file enroute IFR and shoot VOR and localizer approaches.

If I added to that an audio panel with marker beacon receivers and an external CDI to a slightly upgraded nav/comm, I could fly an ILS for only $4,000 more.

So for my $11,000, I could fly IFR legs of just under 400 miles by myself, having no weather detection, no autopilot, no IFR GPS, no DME, no ADF, and no seat next to me to stow my charts, snacks, and other paraphernalia. I'd be lucky to make 110 knots, but it would cost me only $50 an hour. Adding weather capability would be relatively easy, but at the cost of a few thousand more dollars and a subscription to XM WX.

I would be limited to short-range flights, flying alone or with one passenger, and in relatively benign weather. And the lack of real backups for anything made the whole issue a little murkier.

I decided it made more sense to spend $150 to $200 an hour to rent a comfortable retract single or light twin that's more in tune with the mission involved. But there are a few things that make this decision less than perfect.

First is that rental aircraft tend to be a bit battered. Even the newer airplanes somehow give off an air of tiredness, perhaps because of the legions of pilots who don't care enough to keep their shoes off the seats and sidewalls, or clean their trash out when they land. The older ones smell like sweat, which is okay when it's yours but a little appalling when it's someone else's.

The real thing that makes me regret the decision is

that I love the look on the ramp hands' faces when you taxi a brightly painted taildragger to the apron of a busy metro FBO in cruddy weather.

"We don't get many of those in here," one ramp hand told me after I ducked in under falling ceilings at one of the busiest airports in the Southeast. Even the controllers had been confused, turning "Citabria" into "Experimental."

SENSIBLE ALTERNATIVES II

Every year around the holidays, millions of people traipse off to the homes of friends and relatives for Thanksgiving, Christmas, or New Year's. Lots of them fly. Some of them fly themselves.

One year, my family would likely have been among those flying themselves, but for the Lycoming AD that grounded turbocharged engines of 300 hp or more. So instead, we decided to pack the kids and the dog into the SUV and join the hoards on the roadways from Orlando to Atlanta.

By airliner, the trip was slightly more attractive than in times past because the airlines' woes had kept the lid on fares to some extent. Checking reservations late in October, I could get five tickets on either of two airlines for a couple hundred apiece. Total cost for the five of us: $900–$1,000, depending on departure time.

But we'd need to show up at the airport early, go through the demeaning sweeps that allege to provide security, wait for the airplane, wait for the pushback, make the flight, wait for the bags, then make the hour-plus drive from the airport to grandma's house. Total time door to door: a minimum of five hours, and that's if the flight is on time.

A local FBO had a newish Cessna T206 on the rental line, which isn't fast but carries the requisite load. By that airplane, the trip stacked up this way: about 40 minutes from home to having the bags loaded in the airplane, three hours of flying, twenty minutes to get the kids to grandma's house. Total travel time: four hours. Total cost for renting the airplane round-trip: $900.

Clearly the 140-knot Cessna was the better travel device than the 450-knot Boeing for this trip, assuming the weather didn't have anything nasty in mind that the Cessna couldn't handle. Alas, the woes of the big Lycs grounded the airplane and meant our trip would actually stack up this way: sit in the car for eight hours. Cash outlay round-trip: about $150 for gas and a meal along the way, plus wear and tear on the car, but who counts that?

General aviation has gotten a bad rap in recent years, being painted as either a magnet for crazy terrorists or a country club for the wealthy executives demanding bailouts. However, as these numbers show, small planes can make sense from both a time and financial standpoint, unless they're grounded by weather or mechanical troubles, of course. But the same could be said for virtually any form of transportation.

ANSWERS FOR EVERYTHING

The arrival of spring heralds an important change in the attitudes of pilots. No, we're not talking about the sudden and uncontrollable urge to wax wings and fuselage, but rather the start of airshow season.

Airshows are great. You can gawk at airplanes you'll never be able to afford, check out some you'd never be allowed to fly, watch pilots demonstrate things that in any other context would only be described as foolish. Yet for all their wonder, airshows often hold an attribute most attendees routinely ignore: education.

A pilgrimage to Sun 'n' Fun in Lakeland, Fla., or AirVenture in Oshkosh, Wisc., cannot be considered complete unless you check out the wide variety of seminars and presentations put on by safety organizations, type clubs, manufacturers, alphabet groups, the FAA, NASA, and just about any other organization allowed on the field.

Broad programs focusing on engine operation or spatial disorientation are worthwhile, to be sure, but don't limit yourself to formalized seminars. Avail yourself of the expertise that sits in the exhibitor booths, and you may be surprised at what you find.

For example, one former airplane of mine had installed a Frankenstein autopilot. Combine this unit with that servo with this add-on and pop off into IMC. It appeared to work flawlessly, but the documentation was incomplete when we bought the airplane, and I never really knew how to operate some of the features I knew must be in there.

My partners and a few instructors flew the thing repeatedly trying to figure it out, but to no avail. Enter an airshow. I consulted someone working in the booth of the manufacturer of one of the components. I described what I had and what my problem was. He quickly diagnosed the situation and told me how to deal with it. On the flight home, I tested his advice, and it was flawless.

But that's not the end of the story. Two weeks later a package arrived containing complete documentation for my system, even though it involved equipment from other manufacturers, and a letter of thanks from the representative for using their product. No charge.

These are the kinds of issues that make airshows valuable beyond compare. If you just want to do some load-testing on your Visa card, you can do that over the Internet or at a local pilot shop. Airshows have vast resources at your disposal for just the price of entry. Sure, try on the new ANR headsets. Watch Sean Tucker work his magic. Then show how smart you are and go get smarter.

MORBID INTEREST

The miracle that is the Internet is, as everyone knows, a double-edged sword. For many years I would routinely set aside two or three days every month to analyze every single preliminary aviation accident report posted the previous month on the National Transportation Safety Board's website.

It was morbid. It was depressing. It put me in a foul mood. My family hated these days as much as I did. Some months were better than others. Sometimes there would even be a day when there were no reported accidents anywhere in the United States, and that would cheer me up slightly. Unfortunately, those days don't happen very often—once a month, maybe.

In the six years that I subjected myself to this routine, I became jaded and insensitive. I called people idiots, and worse, for transgressions such as pushing a long landing instead of going around and winding up in the trees. Somehow, I would hope to have more compassion for the misfortune of others.

Perhaps what bothered me the most is that the accident record offers incontrovertible proof that this flying thing can be a risky venture. Few people plan to put themselves in a hospital or a friend into a grave, but it happens with astonishing regularity.

One month, as the D-Days approached, I noticed something different. When I decided to take a sunset flight by myself on a beautiful July evening, my wife called me when I was at the airport preparing to leave. She told me to be careful.

The flight was a slice of serenity in that poetic sort of way that people try to invoke when they talk about the romance of flying. It was like a week's vacation, boiled down into a concentrated hour and a half dose.

When I landed and had cleaned off the bugs, I noticed a voicemail on my cell phone, another call from my wife. The concern in her voice was palpable.

I asked her about it later that night. Yes, she said, she worries when I fly, particularly the Citabria, even though she knows the pains I take to be a cautious and safe pilot. She also inquires about the experience and attitudes of anyone with whom I choose to fly.

I mentioned to her that in the summer there are perhaps 180 to 250 accidents nationwide each month, and her eyes widened. I hastened to add that the majority were minor things such as runway overruns and hard landings that did not cause injury, but that didn't seem to help. I dug the hole a bit deeper by adding "but only forty or so are fatal."

"That many?" Ever see iciness and panic struggling with each other?

"Well, that's in a bad month," I lied. "Usually it's not so high."

She is not a reluctant flier. She accepts the utility of small airplanes even though she refuses to go flying just for fun. Even so, you can imagine how this conversation progressed from there. Unfortunately, it's the kind of discussion many general aviation pilots have with themselves, their significant others, and their colleagues far too often.

What we do affects our safety. But the perception of safety held by those who don't fly, as well as those who ought to know better, is shaped by every careless mistake or lousy decision that happens to find its way into the morning newspaper.

Although it's impossible to supply a one-size-fits-all answer to the question of whether flying is safe, it does no one any good to fall back mumbling something about the drive to the airport being the most dangerous part of the trip.

We face risk by accepting it, understanding it, and taking steps to minimize it. But we need to be active and vocal in our pursuit of safety. After all, if flying is perceived by others as being a foolish undertaking, then at what point do we act the fool by pursuing it?

THE "EASY" BUTTON

The dilemma in which general aviation finds itself is that it is an expensive pursuit in a time of increasing complexity. A shrinking portion of the population is willing to invest the time, money, and effort required to fly airplanes.

In an effort to combat the decline, the industry has trotted out advertising, saying flying is "easier than you might think." Going a step further, NASA's SATS program aimed to create a generation of advanced airplanes that would allow cost-effective transport between cities; that program hinged on getting more people to fly.

On a more personal level, think about how you respond when someone expresses an interest in learning to fly and asks if it's easy. The dilemma, of course, is that you want it to be appealing rather than academic; you want your passion to infect others. However, I wonder if it's smart to take the position that any idiot can learn to fly.

In a broader sense, is convincing non-pilots that flying is "easy" a good thing?

Be A Pilot, an industry organization created to recruit student pilots, produced some of the television commercials in question. Former President Drew Steketee thinks the industry is nowhere near scraping the bottom of the barrel with respect to student pilots. In addition, he says, promoters are getting more sophisticated about those to whom the messages are targeted.

Aircraft manufacturers are not lost in this equation either. Interiors are becoming more car-like, even at the expense of useful load. Hopefully that's an attempt to appeal to non-flying spouses when the flying spouse suggests a new airplane.

I would hate to think it was an attempt to position airplanes as flying cars, because the last thing aviation needs is for pilots to adopt an automobile mentality about things that fly. Take a look around a Wal-Mart parking lot, and you'll see things that make even the rattiest airplanes look positively loved.

So if flying isn't easy, what is it?

Be honest with yourself for a minute. Isn't the challenge part of what got you here in the first place? Don't you feel just a little swelling of pride when new acquaintances become somewhat agog when they hear of your craft?

We need to get off the kick that flying is so easy a caveman can do it. Sure, some parts of flying are relatively simple. But becoming a well-rounded pilot proficient in all areas of airplane operation is a fairly difficult thing to attain and even harder to maintain over the long haul. To steal from one of my favorite movies, "Of course it's hard. It's supposed to be hard. If it was easy, everyone would do it."

We'll see how that plays out at the next cocktail party.

RISK, PRACTICALLY SPEAKING

You know how to fly. There may be no question about that. But that doesn't mean you're a safe pilot.

So many things influence the safety of a flight. Weather. Airplane. Pilot. Maintenance. Luck. The goal, no surprises here, is to stack as many things in your favor as possible. It's a lot more than that, though. It's also important to recognize the circumstances under which you might tend to ignore danger signals.

The biggest failure of general aviation flight training—from initial on—is the lack of emphasis on decision-making and risk management. The FAA's Aeronautical Decision Making model is a feeble effort that focuses on rote memorization of bureaucratese. Clearly the answer is a real-world approach that helps pilots learn where dangers lie and how threatening situations can overwhelm even a prepared, proficient pilot.

In recent years, training has become more realistic, beginning with the King Schools' course, Practical Risk Management for Pilots, and continuing on with scenario-based training offered as part of the transition training provided by some manufacturers. These programs typically create a set of illustrations that outline just where threats to your safety can pop up.

Generally speaking, though, risk management is a neglected aspect of flight training. Primary among the problems is that external pressures have far more influence over a pilot's willingness to embark on a risky course of action than most are willing to admit.

Yet external pressures are given only passing emphasis in most conventional flight training programs. They're consigned to brief discussions of "get-home-itis" and possibly a reference to VFR into IMC. In what appears to be an astounding lack of foresight, the FAA's commercial certificate practical standards contain little on decision-making and risk-analysis skills. This for a license to fly for hire.

In the interest of safety, it's important for pilots to hold themselves to a higher standard than the legal minimum. Most pilots admit this when it comes to instrument proficiency yet seem to think it's perfectly okay to ignore those aspects of training the government has refused to mandate.

I think there are three things every pilot should do at least once. The first is spin training. The second is semiannual instrument proficiency checks for anyone who flies infrequent IFR. The third is to study and understand risk management in all of its forms.

Meeting the legal minimums for any task, rating, or proficiency is like getting a C in flying. You don't accept that on your children's report cards, and you shouldn't accept that in your logbook.

I've flown with pilots who showed superb stick and

rudder skills but decision-making skills that would make any Hollywood bad boy cringe. Unfortunately, current flight training strategies churn them out like license plates.

Do yourself a favor. Demand more of yourself than a C.

FEED YOUR BRAIN

I'm sick of gadgets. I'm sick of browsing through the Sporty's catalog and seeing 500,000 things that every pilot "needs" in his or her flight bag. I'm sick of looking in the local pilot shop and seeing office supplies with a pair of wings silkscreened on them and the price quadrupled.

I get tired of student pilots asking what handheld transceiver they should buy, when what they really need is another hour of crosswind landing practice. But what bugs me most is the apparent conviction pilots seem to exhibit that one more gadget will somehow guarantee the safety of their next flight.

The assets a pilot can carry that enhance the potential for a safe flight are three legs of a stool: information, skill, and judgment. It seems like curmudgeonly pilots emphasize skill and judgment, while the whippersnappers focus on information, with a passing nod toward skill.

Gross generalizations, sure, but how else do you explain the popularity of handhelds among renter pilots who can't use the thing outside the traffic pattern anyway because there's no external antenna? This is a safety item?

Before latching onto any gadget, you have to ask yourself if it's going to enhance the safety or comfort or convenience of the flight in a realistic way. Too often they address the perception of safety rather than a bona fide safety issue.

Take portable traffic detectors, for example. Buyers genuflect in front of them with credit card outstretched, without stopping to realize that midair collisions are virtually non-existent outside of traffic patterns, and in traffic patterns the darn things are squawking so much you can't get any useful information out of them anyway.

But yet, when you're cruising along you will occasionally get notification of an airplane you didn't see but that wouldn't have required any action on your part anyway. Whew. That's a relief.

Now there are some gadgets that represent a safety advance, in my opinion. I'd count handheld GPS/XM weather units and quality headsets among them. But little wheels that tell you what kind of entry to a holding pattern or a level to tell you if you'll be in that cloud ahead? Please.

I'd like to see pilots take the weight they devote to gadgets and put that much more fuel in the tanks. I'd like them to spend the money they devote to gadgets on training or airplane maintenance or introducing a potential new pilot to the world of flying.

I'd like to see them concentrating on the airplane and the instruments or the view out the window instead of being heads-down trying to figure out how to turn on the gizmo that will add useless nuggets of data to their already-occupied brains.

Every year, air show season has throngs of people gravitating to those big vendor-filled hangars to get respite from the sun (or rain). They need to ask themselves if not buying anything might give them the time and motivation to prop up those other legs on the stool.

OLD CHALLENGE, NEW OPPORTUNITY

Leafing through my logbook, I found I was overdue for an instrument check. Well, I already knew I was overdue; I simply hadn't realized just how overdue I was.

Once upon a time, I would have one every six months, regardless of how much time I'd logged in IMC or how many approaches I had made during the previous half year. I continually challenged myself to fly more precisely to be better prepared, and I had an instructor friend who shared my attitude that an IPC is not a signature in a logbook. An instrument refresher—a proper one, anyway—is a demonstration that you have what it takes to fly your airplane on instruments when the chips are down.

You're not flying for your instructor, nor for the FAA. You're flying for yourself, your family, your passengers, and their families. Maybe your company too. But that's pretty far down the list, in my book.

But I digress.

Living in a new city, I went down to the local flight school and talked to the chief pilot. There, I heard the cry that's familiar to Part 91 operators: It's so hard to find and keep good instructors because just about the time a novice instructor (especially a multi-engine instructor) starts getting good, a more attractive career path opens up.

If you are not lucky, you will get a relatively inexperienced instructor who, noticing that you are not a primary student, will tend to defer to your greater experience. Sometimes you end up paying the hourly rate to teach the instructor something. If you are lucky, you will get a sharp-eyed instructor who knows when to shut up and let you dig your hole deeper.

Those pilots who are compelled through emotion or corporate policy or insurance company mandate to attend a refresher clinic at a company, such as SimCom or FlightSafety, have less of a problem with the issue of rookie flight instructors, but still everything may not be rosy. The problem here is how to merge the simulator experience with the peculiarities of your airplane.

Primarily this is an avionics question, although there may be other differences, depending on the vintage and model of your airplane. It's not enough to know how the airplane works and how the avionics work. You also have to know how everything works, or doesn't work, as a package. Gestalt at its finest.

In my case, the situation was not so dramatic. I was able to locate an instructor with like mind. In other words, I made it clear to the flight school's chief pilot that I was not there for a signature but for a refresher. He paired me with an instructor he characterized as a "procedures freak."

So far, so good. The instructor and I meshed fairly well, although he did show that lack of real-world application of technique that is so common among instructors who have spent their time continually flying the same local instrument approaches instead of reaching out into the unknown.

The bottom line was that he showed me some things; I showed him some things. Everyone went home happy.

That done, I had more work to do. I had a plan to buy a new (to me) airplane, a decision that encompassed a couple of airplanes I liked on paper and by reputation but which I had never flown. I talked to long-term owners of the models in question and had a pretty good handle on important modifications, pre-purchase "gotchas," good-to-haves, and those items that dictate walking away from the deal. The next task at hand was to get some yoke time to help me weigh the different models accurately and line up some transition training.

Meanwhile, I continued to do the fun stuff—shopping for airplanes. I identified a couple of airplanes that looked like winners on paper. The prices on light twins were tanking seemingly on a daily basis, and I figured each day I didn't buy one saved me a few bucks. My gut feeling at the time was that the market was probably pretty close to the bottom. Knowing that you make money on airplanes when you buy them, not when you sell them, I continued my search for the right airplane at the right time.

While all that was going on, I found myself brushing up on instrument skills that had somewhat atrophied while I was playing with my Pitts. But a new airplane means both a new challenge and new opportunities. And who can argue with that?

CREATING USED AIRPLANES

We've said it to ourselves. We've said it to our spouses. We've said it to our colleagues, our customers and our accountants. If not, we've thought it.

When you're asked why the airplane does (or doesn't) cost so much, carry more fuel, go faster, have a potty, get delayed by weather, hold more people, or require recurrent training, how do you respond? My guess is it's some variant of the time-tested truism: All airplanes are compromises.

You can buy speed. You can save money. You can get a spacious cabin. You can use a short runway. If you want to do all four at the same time, then I guess it's time to start thinking about your priorities.

The history of aviation is littered with airplanes the marketing department implied could be all things to all people. That didn't turn out to be true, of course, because the "best airplane" means different things to different people.

Did we learn? Even now, manufacturers are hawking "sliced bread" airplanes. Whether they succeed or fail depends on a lot of factors: real life performance, reliability, cost, the financial strength of the company, fuel prices, hardware, insurance, safety, ergonomics... the list goes on.

"So what?" you're asking yourself. So plenty. Aircraft designers and the manufacturers who bring those designs to life matter to you, regardless of whether you're in the market for a new airplane. You see, odds are you're not the only one with similar needs and preferences, and maybe some of those people are ready to buy new. Someone has to create used airplanes for others to buy, so these early adopters are essentially creating the airplane you may be flying next decade.

The question you have to ask yourself is whether the airplanes being designed and built *now* are things you will want to fly then. If so, good for you. Watch your investments and tailor your experience to make it happen. If not, don't sit idly by and watch as others define the market for you.

No, now is the time for you to approach manufacturers and let them know what kind of airplane you'd like to be flying. If you don't, you'll have no option at that future date other than to settle for flying someone else's dream.

For example, as vintage designs age, the newer examples tend to gain weight from added insulation, nicer upholstery, and burgeoning options, thereby cutting into your useful load. Sometimes the engines can be up rated to keep the performance up, but that comes at the expense of fuel burn, range, and sometimes handling.

The upshot is that if you are concerned about useful load, tell the guys who make the airplanes. If you don't give a whit about fuel burn and want every knot you can buy, mention that too. Customer feedback is a sincere form of market research, and that applies to potential customers as well as existing ones. With production volumes as low as they are, most manufacturers welcome whatever prospects they can get.

Without guidance from buyers or potential buyers, the manufacturers will tend to build what the dealers ask for. And anyone who has bought or leased a car knows what that means—options bump up dealer profits. Not such a big deal in cars, but a useful load hog in airplanes.

The same goes for other manufacturers. If, for example, you want a certified solid-state replacement for your gyroscopic attitude indicator without springing for a total panel rebuild, you're out of luck. Sure you can get a more comprehensive glass panel refit, but maybe that exceeds your needs and your budget.

Unless the manufacturers know buyers are interested in an AHRS-powered artificial horizon that slides into a 3-inch hole, they're not likely to go to the trouble of building and certifying it; squeaky wheel and all.

Look at the experimental market, and you may find a number of products you'd like to install in your Cessna 340, but you can't because of certification rules. Speak up. Tell the manufacturer you'll buy it if they certify it. Without that feedback, few will go to the trouble and expense.

The next time you're at an air show like Oshkosh or Sun 'n' Fun or a convention like NBAA or AOPA or at a type club convention where manufacturers are represented, stop by and tell them what you want. It may take a decade, but someday you may be glad you did.

THE RUINOUS ROAD

Every time a holiday travel period comes and goes, my mind wanders to the notion of how airplanes and cars serve much the same purpose: getting us and our stuff from one place to another as efficiently and comfortably as possible. However, such introspection raises something we all know on a gut level—how the two modes of transport differ—to a level of awareness that makes it all too easy to become a cynical snob with regards to those poor pedestrians we call non-pilots.

For me, a recent holiday season meant some time on the road, as my airplane sold in November and my search for another had not yet born fruit. (Well, plus we took a trip on which we wanted to bring the boat with us.) Maybe I'm starting to get those nostalgia flashes that I've heard start creeping up when you hit middle age, but I can't be the only one who finds the disparity between traveling by car and airplane growing ever wider.

Fill in your own blanks, but every time I launch in the Honda, some nagging questions come to mind:

- When did "speed limit" start meaning "minimum acceptable speed"?

- Does anyone hit their brakes when the light turns yellow anymore?
- Why do some people seem willing to risk an accident to avoid being one car length farther back in a line of traffic?
- Anybody else noticed the tendency of some drivers to launch into road rage mode if someone else does *anything* that makes them slow down or otherwise feel the slightest bit inconvenienced?

Compare that to flying. The regulatory low-altitude speed limit is generally not a factor that is ignored, and for many aircraft owners something of a pipe dream anyway. Traffic and traffic jams are impersonal, relegated to IFR releases, landing slots, and other inconveniences that generally do not involve eye contact with other travelers. Perhaps the sole exceptions are arriving at a crowded ramp or trying to use the self-serve fuel pumps. In either case, relations are generally more cordial than you'd find at an urban interstate highway interchange at 6:32 p.m. on a weekday.

Wait a sec, you're saying, it just isn't fair to compare the two, especially given the training demands, cost, and specialized usage involved in anything aviation. Automobiles, however, are an integrated part of society that's more egalitarian than the world of airplanes.

True and true, but aviation is not without its share of frustrations. Maybe you've been fortunate enough to land at a small airport at night and find that not only was it unattended, but all of the gates that could let you *out* of the ramp area were locked. I recall one winter night on the ramp in Crossville, Tennessee, when I was induced to stay in the airplane burning avgas while waiting for a relative to arrive or risk freezing. Other pilots tell stories of climbing fences and sometimes snagging clothes on barbed wire.

Other frustrations include being stuck at an FBO for several hours when your attitude indicator takes a powder as you taxi for departure on an IFR trip. And so it's on down to the local diner for a meal you didn't really want while a couple hard-working mechanics install a loaner that will get you home. You can insert your own batch of woes in here, I'm sure.

The point is that it's not the frustrations that matter but how you handle them. I dare say I haven't seen many pilots flipping off the pilots of other airplanes just because the ramp is full. And what about sticking your wingtip out from taxiway Alpha 4 into taxiway Bravo so that pesky airplane can't get in front of you?

Courtesy? Perhaps. But I'd like to think it's more than that. Maybe pilots live in a world where we all sort of feel like we're playing on the same team, regardless of how many logbooks we've filled up or what kind of fuel our engines burn. I imagine it was like that in the early days of automobiles, when the drivers were dodging horse apples instead of potholes, and the automotive fraternity did not include a sense of entitlement.

How you deal with the situation when something goes wrong is a big part of being a pilot. Too bad it doesn't seem to be something we can teach to the general public.

THE MISTAKES OF OTHERS

As a student of general aviation safety, I have always been frustrated and discouraged by the inability of pilots to learn from the mistakes of those who have been there before.

Month upon month, year upon year, pilots crash for a distressingly small range of reasons. A handful of causes such as fuel exhaustion, runway loss of control, controlled flight into terrain, buzzing and VFR into IMC, account for more than three-fourths of general aviation accidents. The obvious question: When will we learn?

Even though we perceive accidents as something that always happens to the other guy, there are several reasons why you should care. First, it's just a darn shame that pilots make boneheaded moves that take innocent passengers with them. That just makes me angry. Second, crashes cost us all money—in the form of increased insurance rates and tighter checkout/proficiency requirements—and credibility due to the perception that GA flying is just for a bunch of reckless daredevils.

And it's not just crashes. Since the use of temporary flight restrictions ballooned in 2001, airspace incursions—

and the resulting huge media attention—have cost the aviation industry allies in the media and government, as well as among the general public. Presidential TFR violations especially carry with them high visibility and the potential to do more damage than an F5 tornado in Oshkosh the last week in July.

I mean, how hard is it to get a briefing? While the privatization of Flight Service Stations has not been without its performance issues, it hardly justifies taking a cavalier attitude toward amassing suitable preflight information.

One recent Saturday I overheard Miami Center making a blind call to an aircraft that went something like, "Aircraft five miles southeast of Gainesville at 3,600 feet, if you're on this frequency, please respond. You've violated a TFR."

Now, on board with me were two passengers who had never before flown in a general aviation aircraft, one of whom was reluctant in the first place to get on any airplane smaller than a 737. The sternness in the controller's voice was unmistakable, and so to hopefully defuse any anxiety the call might have produced, I jokingly sang into the intercom, "Somebody's gonna get in trouble."

My wife, sitting in the back with one of the passengers, asked if the controller was talking to us. Silly girl. We were nowhere near 3,600 feet or Gainesville, and we were on an IFR flight plan. Later I realized the question was for the passengers' sake, but I missed it at the time.

After three of these calls, a voice came on that I'd not heard on frequency to that point. The pilot did not identify himself or his airplane. His radio technique was atrocious, with the rambling, pause-filled conversational nature that so frustrates pilots and controllers who want to use frequency time to actually accomplish something.

"Center, just out of curiosity, what is the TFR?" the pilot asked. (Somehow I doubted his question was simply a matter of curiosity.)

"It's for an air show," came the controller's reply.

"For what?"

"An air show at Gainesville."

"Oh, yeah, I think I heard something about that."

Want to put money on whether this guy was the offending pilot? Unfortunately, I got switched to Jax Approach and did not hear the outcome of the conversation, but I can only hope it included a request to IDENT and then "call this number."

If that sounds intolerant, it is. I will admit to having made my share of mistakes in more than twenty-five years of flying, and maybe one or two would have ended with a bust if I'd been unlucky. But never have I flown without making a conscious effort to do things the best way I knew how.

Preventing accidents and staying out of TFR airspace have something in common: an attitude that bad things *can* happen, but I'll do my best to make sure they *don't* happen today. While almost no one goes out planning to crash, the effect of poor preparation is nearly the same. Too many pilots seem to have their heads stuck in the sand, firm in the conviction that their skills or their hardware will get them out of a jam. Too bad many of those jams could be avoided in the first place if we'd only learn from the mistakes of others.

AIRLINES VS. GA

Like a lot of pilots, I smugly pretend flying my airplane for business and pleasure trips beats the airlines hands down in all respects, with the exception of cost per mile if I'm flying alone or with one other person.

Sometimes I'm right, and sometimes I'm wrong. There. I said it.

Two recent trips underscored how sometimes you win and sometimes you lose. And while these flights are hardly unique, I think it's interesting to examine them, even if I may be accused of preaching to the choir by doing so.

Trip number one was a planned flight from Jacksonville, Fla. to Waco, Texas. Because I fly a slow aerial minivan rather than a fast one, and because it was ever-so-slightly farther than my comfortable nonstop range, door-to-door time would be nearly seven hours each way, depending on how long the fuel stop took.

I was planning to fly it; however, it dawned on me that I would not be at my best at my same-day appointment if I were on the tail end of a seven-hour trip. I investigated airline flights.

When booking airlines for business, I am often torn

between expensive refundable tickets and cheap nonrefundable ones. While I tend to lean toward the cheap ones, I've gotten nailed a few times with schedule changes that have had me eating rebooking fees (or entire tickets). This schedule looked solid, and I was able to find a route on Southwest Airlines that took me into Dallas, where I'd rent a car and drive the rest of the way. The affordable fare was even refundable.

The resulting trip would take almost as much time, but I decided I'd be better off relaxing in the back of the plane and only driving an hour and a half. In either case, I'd overnight in Waco and come back the next day, so schedule there was not a factor. There was one scare, however. I arrived at the departure airport to find the security line ten times longer than I'd ever seen it before. I got to the gate just before they shut the door for departure.

Score one for the airlines. That trip went courtesy of Southwest Airlines. Airlines 1, GA 0.

The next trip was from Jacksonville to Oshkosh. One stop in my airplane; change planes in Detroit, if I flew Northwest Airlines. No-wind, door-to-door time came out as 6.5 hours in the airplane, again depending on the length of fuel stop. By airline, about six hours.

This time, my arrival time was more flexible, allowing me to cope with headwinds, weather deviations, or an unusually long fuel stop. And allowing my departure to be flexible helped because with Oshkosh, you never know when you might want to stay another day.

As it turned out, the weather over Detroit, my connection city had I gone airline, was atrocious the day I was returning. In my airplane, I opted to fly the southern route west of Chicago rather than heading over the lake and down through Michigan. By airline, I would have sat in the terminal at Appleton, Wisc., or Detroit for who knows how long. By GA, I was home in time for dinner. That evened up the score, GA 1, Airlines 1.

Note that I have left cost out of this analysis. Anyone who flies himself around has come to terms with the cost of flying. Rationalize it, ignore it or embrace it, we all have our ways of coping. Yes, for both trips the airline option was less expensive, but if it were all about cost, very few of us would play this game in the first place.

As a side note, at Oshkosh I ran into an old friend who is exploring expanding his business to include aviation companies. While I'd taken him on recreational flights years ago, only now is he learning firsthand the convenience of general aviation. I smiled knowingly at his epiphany, because it's something I've been telling him for years. But now he knows for himself.

Will it lead him to begin flying himself instead of hooking rides from in-the-know colleagues? I have to believe it might. He travels a lot and has seen the airlines drop the ball countless times. So far his exposure to general aviation has been limited, but GA still has a perfect record in his book.

It won't stay that way, of course, but over the long run, I think he'll find GA will be ahead of the game, as long as there's money to pay for it.

COLOR ME GREEN

Green is the color of money, and perhaps not coincidentally it's also the color of envy. Of late, however, green has taken on a new importance as public emphasis focuses on environmental causes.

Personal aircraft are not really a tree-hugger's dream, but neither are they an environmental nightmare. There simply are not enough general aviation flights to come close to the real target: growing airline traffic worldwide.

Aviation as a whole generates about two percent of global manmade carbon dioxide emissions, but airline traffic is growing rapidly, and jet engine exhaust goes directly into the upper atmosphere, where it can't be mitigated as easily as ground emissions. In addition, the water vapor in the exhaust may have three to four times the warming impact of the carbon dioxide in the exhaust.

Whether you agree with the global warming premise or not, it's a public relations problem of the highest order.

Curiously, when it comes to airplanes, money, and environmentalism come together through the cost of fuel. High fuel prices push pilots to work harder for the most fuel-efficient altitudes for their aircraft. They may bring the thrust lever back a touch. Owners may upgrade older engines to more modern, fuel-efficient ones; they may even switch to more fuel-efficient aircraft.

Other fuel-saving strategies might include working harder with ATC to negotiate direct routing. You might be able to plan your arrival/departure times so as not to conflict with busy times at either your origin or destination.

On the airplane, address any items that might be costing you efficiency. Improper rigging of the flight controls, flaps, or gear doors can cost you a bundle due to increased drag. There may be aerodynamic mods that can boost cruise speed or rate of climb. While these may cost money to do, the operating costs will shrink somewhat, and the payback time will shorten as fuel prices climb.

But if the cost of fuel is really putting a pinch on things, the worst false economy is to cut back on flying so much that proficiency suffers. The safety issues are somewhat obvious, but the performance (and economical) issues can crop up when you least expect them.

Early in my flying career, I owned a Mooney 201. There was a period of nearly six months when I didn't fly because of work and the impending birth of our first child, followed by my wife's recovery from same. On breaking out of this layoff, I took off for a local flight to shake the rust out, and I was surprised by my slow progress over the ground, as the winds aloft were not reported as being that strong.

Only after an embarrassingly long time did it dawn on me that I'd not retracted the landing gear.

While that's an extreme example from a 150-hour pilot, the same thing applies in principle to pilots of all experience levels after any layoff of unusual length. Checklist discipline seems to be the first thing to go, followed quickly by issues that affect the efficiency of your airplane.

Clearly if you are behind the airplane in the first place, it would be unreasonable to expect to operate it at peak efficiency. A better case of "practice makes perfect" would be hard to find. Remaining proficient is the first step in flying efficiently. So far, I've adopted the "suck it up and pay the price" approach when it comes to dealing with higher fuel costs, but clearly there is a point at which that strategy becomes a fool's errand. Looking at the used aircraft market, it appears a fair number of people are downsizing—medium twin to light twin, light twin to heavy single, turboprop twin to turboprop single etc. But for many, if not most, trading down is a difficult thing to do.

Most pilots have, from time to time, resisted the urge to throw up their hands and announce they were fed up with the cost or hassle associated with flying. Judging from the attrition rate, a fair number give in. But for those who remain, the hassles of remaining instrument current pale in comparison to the hassles that await at the airline security line. The pleasure returned by flying on your terms outweighs the angst of another annual inspection. And the cost of fuel, well, it is what it is.

What steps do you take to minimize your fuel costs? Do you fly with a lighter load by having more air in the fuel tanks? Have you adjusted your fly/drive criteria? Do you take commercial flights more often? Charter?

As fuel costs rise, the actions aircraft operators must take escalate from minor to substantial, with "grave" a target we hope not to hit. Because if there comes a time when that small voice saying "give it up" becomes a loud voice, color me green with envy, because the blue skies will surely have turned gray.

ALL GROWN UP NOW

The Sun 'n' Fun air show in Lakeland, Fla. kicks off the flying season for a lot of folks, especially those who use their airplanes for recreation more than for serious transportation. Historically, SNF was Oshkosh Lite, a smaller version of that mother of all air shows in Wisconsin. And like its big brother, it grew up on experimentals and Skyhawks and Pitts Specials.

My, how things have changed.

These days, one corner of the manufacturer display area might be crowded by Citations and other light jets. Nearby might be a couple of TMB850s, a Meridian, and cabin mockups of both the Meridian and the PiperJet, a D-Jet mockup, A King Air C90GTi and a Premier 1A. One year I watched as a horde of Pilatus workers struggled to move the big turboprop single along plywood sheets laid out over 300 feet of rain-soaked mud bog.

The show demonstrated in clear fashion that the GA market is splitting in two directions. On the one hand, the light sport aircraft seem to be taking the piston single end by storm. The other end is obsessed with speed, which means turboprops and jets. Aircraft in the middle, piston single retracts and twins, are being squeezed out.

Interestingly, glass cockpits are the norm on aircraft at each end. Synthetic vision is also poised to become standard fare, along with infrared cameras that can help pilots peer through fog or darkness.

However, the move toward high technology in aircraft is slowed by the long design/certification cycle, which puts moving maps in cockpits that contain electronics that are vastly outdated by consumer products standards. Engine improvements seem to come in fits and starts, and even then the changes are evolutionary rather than revolutionary. And the difference is even more dramatic in airframes. Advances in aerodynamics and materials sometimes show up as aftermarket mods but only for some of the more popular models.

Still, it is the spirit of the experimental movement that drives much of what you see at air shows like Sun 'n' Fun. A modifier might come up with a mod that catches the eye of the manufacturer, as has happened with Blackhawk and Raisbeck coming up with King Air improvements later incorporated by Beech. But for the airplanes no longer in production, and that's really the majority of general aviation airplanes, aftermarket additions by STC are often the way to go.

So in that sense it was appropriate that, tucked away in a relatively quiet location, the owner of a Twin Commander turboprop once had his airplane on display—Sagem glass panel and all. Was the airplane there to show off the mod (and the shop that did it), or was it there because the airplane was for sale?

It's this entrepreneurial bent that lends the spice to

shows like Sun 'n' Fun, and it seems to me that aviation could use a little more entrepreneurial bent these days. Some other random Sun 'n Fun observations:

- Potted plants can hide all sorts of flaws on static display aircraft, especially around the landing gear.
- There are several ways to cope with adversity. I'm pretty sure that just getting ticked off about it is not the most productive route. The massive 2008 mud fest brought out a lot of that, particularly in the parking area.
- The cost of any upgrade connected with a glass panel starts at $9,995, just like speed mods used to be priced at $1,000 per promised knot for piston planes and $5,000 per knot for turbines. The computer designers responsible for glass panels clearly have a handle on "what the market will bear."
- Innovation in aviation is cyclical. One year the news is full of revolutionary products; the next year manufacturers are holding press conferences about paint jobs and flap handles.
- It's no wonder more owners are complaining about aircraft payload. Nowhere is the broadening of America's waistlines more evident than on the food court of a large public venue.
- The larger the airplane, the more likely you are to look back instead of forward when you enter the cabin.
- No matter what you fly, it's always 30 knots too slow.

LITTLE JETS, BIG CHANGES

The economy may be cool, but the world of turbine aircraft development continues to look like a growth industry for the long term as Cirrus, Piper, and Diamond inch ever closer on their single-engine jets.

It's often said the major advances in aircraft development are tied to engine technology, and nowhere is that more true than at the smaller end of the cabin class market. The small Pratt & Whitney and Williams engines are giving the airframe guys something to work with, rather than just think about, and the ultimate beneficiary will be those aircraft owners who are now forced to squeeze every bit of performance possible out of a pair of six-cylinder, air-cooled piston engines.

Diamond, of course, has been flying its D-Jet for a while and delayed its certification program in order to swap to a more powerful engine. Both Cirrus and Piper are flying their small jets, and each program is more or less on target to change the world (if you listen to the companies' press releases, that is).

Embraer brought its Phenom 100 and 300 out on schedule.

Cessna, fresh off the successful development of the Mustang and with a couple big jets already announced, might be looking for another challenge, and we wonder if a single-engine jet might be bouncing around in the minds of its design center engineers.

The low end of the cabin-class market has been dominated for years by the turboprop singles from Pilatus, Socata, and Piper and the venerable King Airs. No pressurized piston twins have been built in two decades (except for a handful of Adam A500s). But the new generation of small turbofans has decent enough fuel specifics that smaller airframes begin to make sense. Turbines aren't cheap, of course, and the economics of personal aircraft lead many owners who would not want a piston single to accept the choice of a single turbine engine.

Performance is part of the issue; however, the continued availability of 100LL is helping the push toward turbines. On the down side, however, the cost of maintenance and turbine-sized fuel burns will complicate matters. As in many things aviation, there just aren't any easy answers.

From time to time it looks like Jet A-burning diesel engines might be the answer for airframes deemed too small for turbine applications. That notion took a hit when Thielert faced bankruptcy in Europe; however, diesel engine programs at other manufacturers stepped in to pick up the slack—after an appropriate delay.

In addition, the success of the small turbine programs at Pratt and Williams may serve to spur the development of even smaller turbofans, particularly if sales of personal jets reach the kind of volume Cirrus, Diamond, and Piper hope.

If that does come to pass, the FAA's current requirement that pilots have a type rating for turbofan operations will begin to look ludicrous. Already a relatively large number of pilots are failing in their attempts to get type-ratings that allow single-pilot operations in Eclipse 500s and Cessna Mustangs. The issue is not speed, which will become even less of an issue with the advent of jets that fly lower and slower. The issue is not complexity, because automation is taking away most of that. The issue, eventually, will boil down to bureaucracy.

While the FAA has done a lot of good over the years in helping to make flying one of the safest things you can do, there is no doubt that it, like any large organization, is handcuffed in its ability to cope with innovation.

When the newest thing out there was a glass panel, that recalcitrance was generally reflected in boxes that were more expensive and slower to get to market than they could have been. (Check the experimental market if you don't believe that.) However, when we're talking engines, the whole system bogs down because of the impact the engine has on airframe design.

The success of any of the new crop of single-engine jets will have an enormous impact on the ability of airframe manufacturers to coax people out of their thirty-year-old piston twins. That is, unless someone comes up with a new crop of small turbine *twins*.

THE TRAINING DILEMMA

Pilots facing the joys of recurrent training have a couple of choices: They can resign themselves to the need to get a signature in their logbook, making the best of a bad situation, or they can embrace the opportunity to dust the cobwebs from the darkened corners of their brains and polish the rust from seldom-used skills.

Although I will admit to on occasion falling victim to the former, I make a conscious effort to do the latter, regardless of the circumstances of the training. From time to time, I'm afraid, the training turns into a game of stump the instructor. Hey, take motivation where you can get it.

Recently I was unlucky enough—or lucky enough, depending on your point of view—to need a flight review. My insurer did not require recurrent training outside of the FAA requirements, and while I'd done a fair amount in the last two years, it wasn't the right kind to fill in the flight review square in my logbook.

There I was sitting across the desk from an experienced flight instructor, who kept drilling down, looking for the chinks in my memory of the FARs, ability to read charts, and other such minutia. Believe me, he found a few. But along the way, I found a few of his as well, and we turned it into a positive experience for both of us.

At that point, it could have escalated into one-upmanship, with each of us focusing more on minutia than the practical, but for once we were adults about such things. Go figure.

The fact remains that far too many pilots approach recurrent training of any kind as an intrusion on their lives. Many flight instructors report a disturbingly large percentage of their clients are experienced pilots and aircraft owners, who ought to know the value of being prepared but who seek the easy way out when it comes time to fill in the legal boxes.

The unfortunate upshot of that attitude is the growing tendency for insurance companies to require recurrent training annually, or just more often, at a facility with a program known to be rigorous. On the face of it, the issue for the insurers is lost history, but I believe in many respects it goes deeper than that, to mere trust.

Some will argue that trust has no place in the business world, but I'm a handshake kind of guy, and I expect there are lots of other people who like to believe that of themselves as well.

Beyond the obvious, it seems to me the schools that make a business out of training people in high-performance airplanes have a compelling argument in favor of using their services. It's similar to the environmental appeal that plug-in electric cars have over hybrids. A plug-in will use power produced at a small number of central power stations that are continuously upgraded to the most modern technology and monitored for proper operation.

The hybrid will be maintained by a greater number of mechanics operating in less-controlled environments. The parallel is rather compelling.

That's not to say that independent flight instructors or small FBO-type schools cannot offer top-notch training. In fact, many do. But just as franchise restaurants are winning out over mom-and-pops, big schools are taking a lot of the high-performance businesses, leaving primary training as pretty much the only fish local flight schools have to fry.

In some cities, the FBOs have decent airplanes available for rent: light twins, Bonanzas, and the occasional Malibu. However, as more of the recurrent flight instruction gets shunted from the local FBOs to the big flight schools, many of those options are going away, which is important if you need a backup airplane when yours is down for maintenance. Increasingly, if you need a fill-in airplane, you will be stuck with a Cessna 182 or Diamond DA-40—decent airplanes for what they are but hardly a desirable temporary replacement for someone used to flying a medium twin.

If that's important to you, vote with your wallet. Throw the local flight schools a bone when you can. Who knows, that may keep your skills sharp enough that your next visit to SimCom or FlightSafety might turn into your chance to play stump the instructor.

And we all know how much fun *that* can be.

BREAKING ROUTINE

There's been so much talk lately about the allure of jets.

It's easy to see that when it comes to performance and operating ease, a small jet has it over a big piston any day. The performance gap between turboprop and jet is smaller, and there are several applications in which turboprops will be around for a while.

To the uninitiated—that is, the general public—a "real" airplane is powered by multiple turbine engines. So far, so good. However, that same general perception also holds that "real" airplanes also feature multiple flight attendants, are loaded by uniformed ramp workers slinging bags, and require various degrading searches prior to boarding.

So if the general public is wrong about what's *in* the jet, could they also be wrong about the jet itself?

Not likely, is your probable response. Fast, smooth, quiet—it's hard to argue with what you can accomplish with a couple of turbines strapped to your tail or wings.

As pilots move up the food chain, "faster and higher" is generally the mantra. With faster and higher, you often get bigger. With bigger, you get more luxurious. Human nature being what it is, pilots tend to look at step-up airplanes with a calculating eye, ever lusting after a few more knots, a few more pounds payload, a little better avionics, a nicer cabin.

Complicating that search, of course, is the penalty in acquisition costs, fuel burn, and maintenance issues that crop up when you step up. For some pilots, proficiency issues and insurability come into play. No matter how you slice it, it isn't cheap and it isn't easy.

All that sounds so cold and clinical. Granted, there are many aircraft owners to whom the machine is simply a tool to get them from here to there quickly, conveniently, and comfortably. In the owner-flown segment, however, I would like to believe a significant percent still enjoys the challenge of flight and the rewards of meeting that challenge time and again.

Who doesn't get pleasure out of a perfectly flown instrument approach to minimums? How rewarding is the look on your passengers' faces when you arrive at your destination hours before an airline connection would have gotten you there? And who—be honest now—doesn't feel even the slightest bit smug when a colleague or acquaintance raises an eyebrow at the news that, well, of *course* you haven't flown the airlines in umpteen years.

More than twenty years after discovering that I really could fly an airplane, I still believe there's more to the flying equation than that whole utilitarian thing, and often I wonder a little bit about that side of the coin.

I'm no Luddite, but I'll admit to a certain soft spot in my heart for taildraggers, especially those with wings on top and below. I find the drone of a propeller airplane en route to the executive airport near my home to be much more appealing than the whine of a jet.

Who can't land at a grass strip, pop the window open, and marvel at the flood of smells that fill the cockpit as you taxi to the little shed that serves as the FBO? Or maybe you've flown formation in an old military trainer with a bunch of other folks flying their old military trainers, following a lead who was simply droning aimlessly along the countryside.

Who can forget the first time they flew solo or performed their first aerobatic maneuver or took off in an airplane they'd built with their own hands? And if you've never splashed down in an isolated lake and bobbed there, engine off, listening to the birds discuss your floatplane, you've missed a tasty slice of life.

There are nearly as many facets of aviation as there are of life, and if you ever feel pigeonholed or bogged down in the hassles that involve remaining a safe and competent PIC, it might be time to rediscover some of those other avenues.

You see, for all the usefulness you can wring out of an airplane, there's still a place for the romance. Despite rising fuel costs and airport closures and NIMBYs who think the poetry of flight is simply poor songwriting, there will always be those whose wings are fanciful as well as practical.

If you're one of them, I applaud you. If you've lost that spark, can I suggest you make another try to ignite the flame again?

SOMETHING OLD, SOMETHING NEW

Pilots are an odd bunch, on the one hand decrying the relative lack of innovation in aviation, and on the other, eschewing new products "until they are proven."

Despite that attitude, new technology does leak into the fleet, through manufacturer improvements, sure, but also through an ambitious aftermarket. While most noticeable in avionics, improvements can also be had in engines, airframes, and even the cabin. Innovation hasn't been as noteworthy as in, say, the auto industry, but we'll take what we can get.

The change in panels, of course, has been huge. When I started flying, most airplanes did not have intercoms, and headset use was relatively rare. "Active noise reduction" meant shouting a little less loudly over the engine noise. GPS was classified. Today's glass panels were the stuff of science fiction.

I routinely am amazed at how downloadable weather fundamentally changes the go/no-go decision during periods of questionable weather. Even after flying with XM weather for several years, I still find myself pinching myself during times of active weather to make sure I'm not dreaming.

Now add to that mix synthetic vision that mixes terrain and obstacle databases with traffic information and displays it on the attitude indicator. Such capability would boggle the minds of the curmudgeons who haunted the quiet GA airports twenty years ago.

On the wings (for some, the nose), the changes have been somewhat more subtle. Engine upgrades have been slow, with accessories showing somewhat more vibrant development. Still, the power plants of yesteryear are still pretty much with us. Turbine engine development has been slightly more productive than attempts at piston engine development, where technology is starting to look really old fashioned.

FADEC? Fay-what? Full authority digital engine control has been around in cars for a generation but has only begun trickling into the world of propeller planes in the last few years. It's long overdue, in my opinion, because issues stemming from power plant mismanagement, such as over temps and poor mixture control, continue to cause accidents and maintenance woes. Unfortunately, this is one primary example of when identifying the problem is much simpler than fixing the problem.

The airframes themselves are the most stalwart—from the OEMs, anyway. The aftermarket crowd has been a little more aggressive.

Engineering wizards around the country have been fairly successful at using aerodynamic mods to make poor airplanes good, good airplanes better, and better airplanes excellent. Those mods often apply to areas where drag builds up at cruise speed, but they also can lower stall speed, improve single-engine handling, or help an engine run cooler or produce more power.

Extensive airframe improvements are expensive to develop, and many companies are reluctant to tackle them given the relatively small number of potential customers. The improvements have to be relatively significant to justify the cost, and as time goes on it gets tougher and tougher to squeeze more water from the stone.

Riding shotgun among all of these challenges is the FAA. The cost of certification simply cannot be ignored, because high costs don't necessarily lead to high prices on the modifications, and buyers have to see a bona fide financial impact in order to take the plunge. That unfortunate reality is why there are so many old twins and turboprops out there with antiquated systems, demanding far more work on the part of the pilot and costing far more to operate and maintain than a modern design of similar capabilities.

In that sense, aviation is the ultimate chicken-and-egg dilemma. New airplanes and airplane mods are expensive because the market is small, and the market is small because they are so expensive. The cost means everything, because high-priced replacements increases the economically feasible lives of assets that were not designed to last this long, with all that implies with respect to safety, innovation, and performance.

At some point, the only way to break the cycle is for aircraft operators to conclude that certain advancements are worth the price of admission. Glass panels have helped spur the market for new airplanes because buyers have concluded glass panels really do represent a significant improvement in the state of the art, and retrofitting a glass panel is a truly enormous undertaking that so far is approved on only a few models.

During hard economic times it's presumptuous to urge others to spend their money freely, but in any economic climate, there are investments that don't make sense, and those that do. Finally, we are at a point when aircraft improvement seems to have a real payoff.

SENSIBLE DESPITE CHAOS

Three conversations I've had in the last week have started out this way.

First: Flown any interesting airplanes lately? Second: What airplane represents a step up in quality and performance from what I'm now flying? Third: Assume a benevolent soul has offered you any airplane you want, including operating costs, for a year. What do you get and why?

Credit crunch or no, it's clear that lots of people afflicted with this wacky bug we call piloting continue to look onward and upward, always trying to go a little farther a little faster. And who can blame them?

The chaos of the last presidential election proved to anyone who cared to look the value of general aviation. Modern national political campaigns simply could not be done without a jet at the campaign's beck and call. I can't imagine Barack Obama conducting a successful campaign via a two-year webinar, regardless of how technologically astute his campaign staff liked to think of themselves and their constituency.

That said, the economic troubles facing the United States have the potential to create far more lasting havoc than bouncing around the balance of your 401K. Already there are signs that companies are scaling back their plans to buy new corporate aircraft, whether that means cutting back on flying or simply keeping current aircraft longer is a case-by-case prospect. But the cumulative effect, given the capital-intensive nature of aircraft manufacturing, is that the crisis would ripple.

While most successful cabin-class designs are picked up by other companies if the original manufacturer goes broke, that doesn't mean the process is painless. The transition makes parts and support problematic, and there's no guarantee that tooling and other items needed to produce spare airframe parts will be easy for the new company to find or convenient to deploy.

Other technical support, notably engineering support to address new kinds of problems, is also hard hit. The new organization doesn't have the institutional knowledge often necessary to deal with problems that crop up. Guess who pays for their learning curve?

Amid the financial mess, the Transportation Security Administration proposed to expand the "security" measures the operators of large general aviation aircraft must follow. As initially unveiled, the new rules would affect only GA aircraft over 12,500 pounds, but they portend additional costs and obligations for all GA operators.

Sure, the "camel's nose under the tent" cliché applies. It's logical to assume that at least some of the requirements eventually will trickle down to smaller aircraft. But even more immediate is the virtual certainty that any costs for programs or infrastructure incurred by FBOs and GA airports as a result of the rule would be "enjoyed" by all GA aircraft operators to some degree.

The sad reality is that it's not getting any easier to employ general aviation aircraft in pursuit of your goals. Flying is getting more complicated. Aircraft operation is getting more expensive. The hassle factor of each seems to be growing exponentially.

Will this lead owner/operators to park the airplane in an effort to mitigate cost and reduce hassle? With the market for used aircraft drying up, does it mean jettisoning higher long-term costs by accepting a capital loss? Obviously an individual's personal picture changes the dynamic of this, but it's also obvious that many are putting new aircraft purchases on hold for a while, hoping for lower fuel prices, credit to become available again, and confidence in a brighter future to shine through. Only in that way can general aviation flourish.

Winston Churchill once said, "It has been said that democracy is the worst form of government except all the others that have been tried." One can alter that to be that general aviation is the worst form of transportation, except for all the others. And that is why we pay five-figure maintenance bills and accept MPG numbers that drop the jaws of Toyota Prius drivers. It's why we submit to somewhat rudimentary "official" medical exams, pore over charts, and consider the annual trek to the simulator facility to be just another detail of living.

At no time is this more obvious than around popular travel days. Millions of earth-bound mortals are locked on sluggish highways that move like a reptile on a cold morning. Miles above, we scarcely notice the season, except perhaps an extra salutation on the radio or some potted poinsettias in the FBO lobby.

Who said there was no place like home for the holidays?

A STUDY IN CONTRASTS

It's easy to get so caught up in worrying about safety that you forget what got you into this flying thing in the first place. And so it was with me on a soft October morning. I had just returned from addressing the annual convention of the Malibu/Mirage Owners and Pilots Association, for which I'd done an analysis of the safety record of the Piper PA-46. While sometimes it seems studying accidents can cast a pall over everything, my concerns paled next to the PA-46 owners I talked to at the convention.

Here are successful professionals who fly what many consider the ultimate incarnation of a single engine airplane. Fast and pressurized, it's built for travel at high altitude. Unfortunately, the demands placed on the piston engines that have been tried in the airframe have been daunting, and both the Continental and Lycoming engines have demonstrated less than stellar reliability.

As a consequence, engine failure is high on the minds of Malibu and Mirage owners, pilots, and even their spouses. Questions about the reliability of the power plants have led some owners to take drastic actions. One highly experienced pilot steadfastly refuses to fly his airplane at night.

Informational forums dealt heavily with engine care and dead-stick landings. They've even worked out the numbers for dead sticking an instrument approach.

Whether a realistic risk assessment or rampant paranoia, this concern about engine failure has some collateral consequences. The runway loss-of-control accidents that seem to plague other models of airplane are conspicuously absent.

Part of it surely stems from the fact that these airplanes are expensive and sophisticated. They tend to be flown by experienced pilots because the inexperienced ones can't get insurance. But they also know the stakes and are willing to pay the price to stack the odds in their favor.

In a crowd of perhaps two hundred people, I asked for a show of hands of how many pilots had commercial, or better, certificates. Ninety percent raised their hands. Perhaps more surprisingly, half of those kept their hands up when asked if they thought they could pass a commercial practical ride in their airplane that afternoon.

That dedication to proficiency is admirable. If it were reflected among pilots of airplanes with a less checkered mechanical record, it seems general aviation accidents would be substantially reduced.

Proficiency doesn't have to come at the end of an instructor's leash either. What you demand of yourself is a huge part of the equation. One reason I like flying taildraggers is because the connection between airplane and pilot seems much more immediate, like riding a motorcycle instead of driving a luxury car.

With such an innate connection, the demands are less like challenges than they are casual conversations. When I returned from the convention, I was greeted by a cool, cloudless morning that called out for me to polish with my skills. Studying accidents can cast a pall over everything, but on this pristine morning I was driven to fly away the funk with a recreational flight in the Citabria.

I washed the airplane first, so then, of course, I needed to fly it to dry it off. I hadn't flown the taildragger in about a month, so it was off to a local uncontrolled field for some landings. Jimmy Buffet was on the CD player, which was piping through a new headset I also wanted to test.

The air was smooth as I slid effortlessly over a lake toward the uncontrolled airport at Leesburg, Fla. I turned the tunes down low so I could keep track of other traffic on the common frequency. As I arrived, the airplanes in the pattern were negotiating a runway change. There was a bizjet waiting for an IFR clearance at the runway that was soon to be made inactive, but everyone agreed to let the bizjet go when he got his clearance. We were a viable self-governing collective.

A few short-field touchdowns were like glass, and my spirits soared on wings that had been spread too seldom of late. I departed Leesburg and went to another airport that boasts a narrow 3,000-foot strip that has an uphill slope at one end. I like this airport, X04, because it calls itself a one hundred-foot-wide runway, but actually there's a strip of pavement about thirty-five feet wide that runs through the middle of two strips of turf.

You can land on the grass. You can land on the asphalt. You can land on the asphalt, deviate into the grass, and back without hurting anything. It's a country airport in the finest sense of low key—under used and very forgiving.

The narrow pavement sharpens landing skills, and I

tried to play the winds and the airplane in a power-off melody that would match the 400-foot landing roll the POH says is possible. On this beautiful day, I actually got close. I landed, stopped, and took off again in the first third of the runway.

All too soon, it was time to head back home. The stick rode effortlessly in my right hand, and it seemed I needed merely to think of a course of action, and the airplane did my bidding.

The controller at Orlando Executive fit me into the pattern without effort or deviation. I rode the rails down to runway 7 and, when the wheels touched, all three did so simultaneously without so much as a squeak. Feet dancing on the rudder pedals, I taxied to the hangar. And just like that, it was over.

While airplanes are a great transportation tool and often carry unmatched utility, it pays to spend a few hours occasionally reconnecting with the primal urges that first led humans to cast their gaze skyward. Even weeks later, I could lull myself to sleep with the memory of those silky landings and the raw power of takeoff. I'd smile, content in knowing that such fulfillment would again be at hand soon.

PART III
USE SOME SENSE

NOBODY'S PERFECT

A pilot I know has never met a reg he didn't want to bust. Well, that's not quite true, but I've flown when he was PIC on three occasions, and each time he's done something that either scared me or got him chewed out by ATC.

He blazed into the traffic pattern at a controlled airport without saying a word on the radio. He busted Class B airspace off the departure end of the runway. He taxied through (not around) a parking area at night at high speed. In short, he flies more like his first flights were in the Alaskan wilderness than at the air force base where he took his training.

His communication skills need work, as the tower controller who chewed him out could attest. His situational awareness is poor in that the Class B incursion (for which he was not cited) happened because he thought he was farther away than he was. And the taxi episode? You'd think his life depended on the thirty seconds he might have shaved off the taxi in his zeal to get home.

We've all had flights when things didn't go, uh, perfectly. But the important thing is what we bring away from those flights.

Many years ago, I was flying some friends on a local flight and got boxed in by thunderstorms. After turning this way and that looking for an out, I was a little unsure where I was until I looked out the window at two o'clock and saw that I was uncomfortably close to the departure end of a 10,000-foot runway at the Class C airport from which I'd come. I shudder at what a 737 would have done to the rented Warrior I was flying.

That flight was a watershed. Since then, situational awareness has been critical to me. Fortunately, the proliferation of moving maps has come to the rescue of many an unwitting pilot. But the digital icons, circles, and course lines don't absolve you of responsibility for knowing your place in the world. I know what my next five steps will be, and I remember what my last five were.

It's impossible to always prepare for any eventuality, because trouble is unpredictable. If we always act as if we'll crash, and we stop thinking about how not to crash, we'll soon stop flying. But knowing the limits of your ability is an important step. Knowing the limits of your equipment is important too.

Safe flying is knowing where trouble is likely to spring up and making sure you're ready to bat it back down. It takes communication, it takes planning, and it takes patience. This is one area in which two out of three isn't close enough.

Ken Ibold

USE THE FORCE?

Look around you at the best pilots you know.

Are they the ones who have 5,000 hours with 4,900 in the make and model they're flying now? Are they the ones with 1,000 hours spread among forty-three different kinds of aircraft?

A primary factor in safe flying is to never assume that you've done it all or seen it all. If every flight contains a lesson, then the level of risk continuously declines, all other factors staying equal. The most we can hope for is that we'll learn the lessons easily rather than catastrophically.

For example, the other day I was in a hurry to depart and started to taxi without releasing the parking brake. I'd gotten a little sloppy about checklists and can't think of a less stressful way to be reminded about their importance.

There are some pilots, however, who don't appear to want to learn. Often in the name of safety, these pilots are so cautious as to be virtually paralyzed. Every action, every decision is carefully deliberated, contemplated, and considered. Is this the kind of thinking that's useful in a time of crisis?

One of the best ways to ensure you keep learning is to broaden your horizons at every opportunity. Check out in a new type of airplane, even if you don't think you'll be able to fly it afterward. Not only will it teach you something about flying your regular plane, but you may find your eyes opened to a whole new world of flying. Vintage airplanes, taildraggers, amphibs, acrobats, twins, helicopters, gliders, light-sport aircraft, and cross-country cruisers are all out there waiting to be discovered. And each one can teach you something about flying you don't already know.

Another tactic for keeping your brain active is to break routine. If you've had pancakes at the same airport restaurant at the same time every Saturday for the last few months, it's long overdue. Fly somewhere for dinner instead. Explore new airports and different parts of the country. Mountain flying teaches you one thing about situational awareness, trying to find the right airport in Florida's dime-a-dozen airport geography demonstrates another.

Finally, you might also try getting a new rating. Instrument, commercial, multi, seaplane, you name it. If you can't afford the money or time commitment a new rating requires, add an endorsement for flying a retractable or high performance plane or simply fly a few hours in an unfamiliar rental.

The quality of a pilot is not determined by the number of pages in the logbook. What's important is the pilot's approach to flying and the willingness to admit that there's a lot more to learn. Next air show you go to, watch a Pitts tumble through the sky and say to yourself, "Gee, that could be me."

DENVER'S LESSON

More than fourteen months after the crash of singer John Denver in a Long-EZ off the California coast, the NTSB released its final report on the accident.

I'd examined in detail the vast majority of accident reports filed in the previous decade, and this accident caught my eye.

The investigation of the last flight of N555JD is noteworthy because Denver's high profile sparked a more thorough investigation than might otherwise have occurred. The final result of the investigation became well known—that Denver likely lost control while turning around to switch fuel tanks using the selector that was mounted behind him.

But the lessons that can be learned from Denver's crash are more lasting than the no-brainer opinion that you shouldn't mount your fuel selector where you can't reach it.

Though Denver's ability to entertain set him apart from the masses, his ability to pilot airplanes was decidedly average. He had a few incidents in his background. He had about 2,700 hours on his private ticket. Like many other people, he had over the years accumulated a multi rating, instrument rating, seaplane rating, and glider rating. A testimony to his wealth is a Learjet type rating.

The Long-EZ was Denver's new toy. The day before the accident, Denver got a "checkout" that involved two touch-and-goes, some slow flight, and a review of the airplane's systems. In order to reach the rudder pedals, Denver put a pillow on the seat back.

The preparation for the flight, or lack thereof, set the stage for disaster. Although he had flown a couple of other Long-EZ demo flights, his preparation for flying a new type was part of the problem. Amateur-built airplanes are distinct from their certified cousins in that they vary widely in terms of construction quality, weight, power, and even handling characteristics. That means each airplane, not just each new type, has to be approached with the same caution a test pilot reserves for an upcoming flight.

I'm no test pilot, but I don't think cruising over the ocean at five hundred feet is quite the right technique.

Much has been made about Denver's supposed problem with alcohol and whether the FAA was really trying to yank his medical. In a way, the alcohol, the controversy over the medical, and the accident all tie into one common thread: pilot attitude.

Flying is fun, but it can't be taken lightly. Don't think beach volleyball, think motorcycle racing.

The most critical piece of equipment on the airplane is between the pilot's ears. When that stops working, it doesn't matter how many hours you have or how fancy your bird is. Gravity will always be trying to drag you down.

DIMINISHING RETURNS

The discussion started out being "is a twin better than a single?" But soon it was branching into the philosophical questions of How much "more" does it take to make something better? When is more actually less? And can you really afford more anyway?

Die-hard twin pilots smugly point to the capabilities of their machines and say it's no contest. The pilots of singles like to say that the second engine does little but lift its own fuel and fall back on old jokes about the second engine only being there to take you to the scene of the crash.

But twins do cost more to operate than a roughly comparable single, and since most general aviation pilots vote with their checkbooks, the popularity contest is long over. Within the world of singles, however, there is great diversity. There is also the law of diminishing returns.

Consider for a second an early '80s vintage Cessna 172 as a journeyman airplane—basic and entry level. A Bonanza has more panache, more class, and a price tag that's twice the size. Do you get twice as much for your money? (I ain't gonna answer that, at least not here.)

Crank it up a notch. A Piper Malibu has a price that's

doubled again and a little more style and elegance than the Bonanza. But the incremental difference has shrunk.

And so it goes up the food chain of airplanes. A great deal more money gets you an airplane that's slightly more capable—and that's true regardless of whether you're going from Skyhawk to Malibu, Meridian to Piaggio, or Mustang to Gulfstream.

What's this got to do with safety? Plenty.

Pilots are always complaining about the cost of flying. Then they go out and spend big bucks on leather interiors for their twenty-year-old airplanes. Sometimes they get so caught up in the T-hangar one-upmanship that they spend more effort making sure they have the nicest plane than they do learning to fly the thing.

I know one couple who traded in their Bonanza for a Baron. The operating costs socked them so hard they couldn't afford to fly anywhere. This is not smart economics, and it's certainly not smart flying. Why bother?

If money's a limiting factor, and it usually is, spend it where it will do some good. Instead of spiffing up the bird and flying with the money left over, fly until you're safe, proficient, and comfortable in the cockpit, and spend the money that's left on the image thing.

A passenger may like the feel of the leather, but that pleasure will be forgotten with the first bounce. Better, I think, to let them deal with a bit of cracked plastic and treat them to a first-class ride. Your polished approach will show you're a person of substance over style.

And who can argue with that when their fanny is hanging out at 17,000 feet?

SELF DELUSION

This just in: If you think you're a hotshot pilot, you're probably not.

A psychology professor at Cornell University found that people who do something badly are usually supremely confident of their ability—more confident, in fact—than people who do it well. Writing in the *Journal of Personality and Social Psychology*, Prof. David Dunning and his associate concluded that the skills required to be competent are the same skills required to recognize that the task is being done well. So someone who can't do something lacks the capacity to realize it.

The researchers cite many examples: people who aren't funny but persist in telling jokes, or stock traders who repeatedly jump into the market and lose out.

We all know people whose brash talk and outspoken hangar flying seem to be at odds with what their flight instructors see come flight review time. Conversely, we also know those who are outstanding stick-and-rudder technicians but who are demure about their ability.

Next time you think you know it all, think of Prof. Dunning and imagine what he might say if you said you hadn't blown an instrument approach in five years or that every landing has been a smooth one.

Flying is unlike most fields of human endeavor because there are few activities, be they hobbies or professions, where the participants spend so much time and effort examining what went wrong. Pilots study accidents, talk about technique, and practice enough to put skiers, boaters, and football players to shame.

But sometimes we do let confidence become cockiness, leading to the attitude that landing minimums only apply to others and that misfortune will dare not cast its hostile glare in our direction. That's when even experienced, conscientious pilots can be bitten.

Don't let anyone get away with pretending they can do more than they can. Even if that person stares at you in the mirror while you brush your teeth.

Avoiding the problem is easy. Here are some suggestions:

- Fly with as many other pilots as you can, even if you have to hire them for a couple of hours.
- Expand your horizons to include new destinations and new airplanes.
- Turn off the autopilot from time to time.
- Let other pilots have their say during those rainy-day hangar flying sessions. You might learn something besides humility.

There's an old song about a stranger in the house who looks like me. Don't let your flying turn you into a pilot everyone else recognizes as the kind to avoid.

Ken Ibold

SHORTCUT ON A LONG PATH

I was planning to take a friend on a short sightseeing trip in a Citabria.

We weren't planning aerobatics, so we skipped the parachutes. I calculated how much fuel we could handle given our weights. After fueling and preflight, my friend climbed into the back of the little taildragger, and I climbed into the front. I started the engine and got the ATIS from the radio. When I called Ground for a taxi clearance, nothing. No sidetone, no carrier, no transmission.

We shut down the Citabria and hauled the Lance out of the hangar. Not as much fun for sightseeing, but even a disappointing day of flying is better than a good day at work.

As soon as I opened the door of the Lance, I felt like Papa Bear must have felt when he got home after Goldilock's visit.

"Someone's been sitting in my chair."

The charts were strewn randomly between the two front seats, and the flight log, fuel tester, and pitot cover were nowhere to be found. An acquaintance had borrowed the airplane and afterward had washed and waxed the airplane and hired someone to steam-clean the interior. I hoped this was his way of saying thanks and not his way of saying, "I just wanted to see if there was an airplane under all that dirt."

I reorganized the cockpit and conducted my preflight. Although the FBO had been instructed to top the tanks, they were substantially less than full. Because the airplane has two linked fuel cells in each wing, it takes some time for fuel to drain from the outboard tank into the inboard tank. Apparently the fueler didn't know that.

Preflight complete, I lit the fire, flipped on the avionics master and was surprised to see things on the panel I didn't recognize. The display on the Garmin had been re-customized to a bunch of settings that seemed less-than-intuitive to me. This was turning out to be a day of unexpected detours.

Garmin reprogrammed, we took off, flew the beaches along Florida's east coast, then looped back around for some passes over what we fondly call "the attractions," those tourist destinations owned by giant entertainment companies that succeed in guzzling your money with the efficiency of a vintage jet warbird.

As we turned to head back for lunch, a small-but-growing thunderstorm stood sentry, blocking our approach path and forcing an end-around.

As we were taxiing back to the hangar, I thought about how this, a short little nothing flight in the scheme of things, could be so full of unexpected twists and turns. The lessons were subtle. Just as expectations can cloud observations, they can also lead to assumptions, that lead to shortcuts, that lead to shortcomings, that lead to trouble.

It's a long path to be sure, but it's one I'd just as soon not start down in the first place.

DEPENDENCY

I just didn't have time.
 Yeah, that was it. I didn't have time in the last few weeks to go practice any approaches or any of the maneuvers research shows could save my life someday, including go-arounds and aborted takeoffs.
 Okay, well there were a few days when I moseyed around the house, not really doing anything constructive. I did watch a few football games that I really had no interest in. So maybe I did have time. But the weather was too nasty.
 Um, except for those few days when I went outside, drew in a deep breath of the crisp air, and thought, *What a fabulous day!*
 Let's see, what other excuses might I have come up with? Didn't have the money to fly because of holiday overspending. That doesn't work. The airplane was broken. Nope, it's running great. I hate flying. Yeah, right.
 As it turns out, pretty much any excuse I can come up with for not occasionally sharpening those tools I may one day depend upon for my life just doesn't wash. Do they work for you?
 In fact, in the last month I have practiced most of those

maneuvers. I did some steep turns with banks in excess of fifty degrees. I shot some practice approaches in perfect weather. I did some chandelles and some go-arounds.

Though most of the practice was just a matter of remaining proficient, I did discover that it's been a while on the go-arounds. In fact, I learned that I had a tendency to raise the gear before the flaps. Don't know where that came from; it just appeared. And few habits could jump up and bite me faster if the chips really were down sometime.

Next week I'll work on aborted takeoffs; it's been a while since I practiced those.

Practicing flight maneuvers is something most pilots forget about. They take their airplanes from place to place and endeavor to be as smooth as possible along the way. If they grease the landing, they feel pretty good about their skills.

Can you think of the last time you intentionally did something in an airplane that you really didn't know how it would turn out? If not, you'd better go fire up that flivver and try it—under controlled circumstances, that is. Just make sure the airplane is capable of it—no hammerheads in a Baron—and if you doubt your skill, bring along someone to keep you safe.

In fact, trying maneuvers you haven't done in a while can be a humbling experience. Maybe that's why we avoid them.

DEPRESSING FAMILIARITY

Much as some people say they like winter flying, there's no way around the fact that, for most pilots, winter only seems to muck things up. Hangar flying replaces cross-country flying, and a whole lot of airplanes get nestled into a maintenance hangar somewhere for an annual inspection.

With spring comes a new commitment to flying, new chances to spread our wings and the reaffirmation of why we took this thing up in the first place. It's easy to get distracted by the prospects of a season full of new adventures and fresh challenges. Maybe a new airplane or another rating is in the works, maybe not. Regardless, this is the time of year to take stock of your flying, yes, but also your attitudes toward flying.

The Air Safety Foundation's Nall Report typically comes out in the spring and often contains encouraging news about flight safety. VFR into IMC accidents have been trending downward in recent years, as have overall fatalities. Mechanical failures don't cause many accidents.

In fact, most years there are not really any surprises in the report, but it does serve to reaffirm the conviction most pilots have that flying light planes is not insanely risky. Recently on one of the internet newsgroups there was a lengthy discussion about risk, with many pilots adamantly contending the statistics don't apply to them because they're too smart/skilled/proficient/cautious to make the mistakes that most commonly cause accidents.

If you believe that, make sure your insurance coverage is adequate. Studies indicate that roughly eighty-five percent of airplane crashes are caused by pilot error. Not pilot stupidity. Not pilot recklessness. Not pilot carelessness. Pilot error.

Most pilots have experienced flights where they've made mistakes that have caused at least a moment of concern (panic?). A little bad luck just then, and maybe bad would have become worse.

Rationalizing away risk does not reduce it. Preparing yourself to make good decisions, armed with sufficient skills and a prudent attitude, is what reduces risk. You can't handle that? Stay on the ground.

Good equipment can reduce risk. That applies not only to well-maintained machinery that doesn't break but also to avionics that add accuracy and capability. The Nall Report says VFR into IMC accidents are down. I've got to believe autopilots, moving maps, and satellite weather have as much to do with that as education and propaganda.

Take a look at the way you make go/no-go decisions and ask yourself if you're truly happy with your approach. Fly enough to stay proficient. Don't just log hours for the sake of saying you've flown. Do something with those hours. And for goodness sake, take along enough avgas. Running on empty remains a primary cause of airplane accidents.

OSH 'TIL YOU DROP

Every year after surviving another AirVenture at Oshkosh, I come home convinced pilots are both the smartest and the dumbest people on earth.

Let's step aside for a moment and try to ignore the pageantry, the forums, the camaraderie, and the sheer amazement that comes with watching Sean Tucker fly an airplane. Instead, consider the Stuff for Sale.

There are few other places on the planet where you could spend so much money and still not get one of everything. Headsets, charts, navaids, engine monitors, weather info, parachutes, hangars, traffic alerts, FADEC, airplanes, engines, seats. That just scratches the surface of the wares hawked at the world's biggest air show.

Pilots are experts at spending money. This makes them smart when they buy the right things and dumb when they misplace their priorities. Oshkosh is a place where they can buy just about any piece of aviation hardware available. Pity that it seems too few ask themselves if they ought to.

If you assume that every pilot's budget has some upper limit, it's evident at AirVenture that most would prefer hardware to software. They'd rather pack their panels than pack their heads.

A forum on upset training and unusual attitudes may boast perhaps thirty people in the audience. A popular one on engine failures will draw hundreds, true, but even that was a small portion of the masses that streamed through the exhibit hangars. The common mantra seemed to be "buy now, learn later."

Oshkosh presents a fabulous opportunity to tap into some of the best minds in the business. The forums contain enough wisdom to save your bacon several times over. Unfortunately, they are overshadowed by all the cool *stuff* there for sale.

A realistic appraisal of the things that can make an average pilot safer has to include learning better aircraft control and practicing instrument flight. Face it—many accidents reflect poor control coordination and VFR into IMC. There is little for sale at OSH that can solve those problems. Fuel flow meters can help reduce the fuel exhaustion accidents, and those are for sale. Engine monitors, seatbelts, and backup instruments are also important safety items, and all can be found in the cavernous exhibit hangars A, B, C, and D.

Still, I can't help but wonder at the sixty-hour-per-year pilots who think a new ANR headset will do more to ensure their continued existence in this world than four hours of practice instrument approaches or crosswind landings. Yet take out the product displays and you lose the crowds. AirVenture becomes just another Wings Weekend at a rural airport.

Perhaps, for one week per year, they should change the airport designator from KOSH to SHOP.

GOTTA GOTTA GOTTA ... SCRUB

The accident chain, like most, was as obvious in retrospect as it was elusive at the time. Consider these links:

Link 1: The pilot and his four passengers absolutely, positively had to be at the destination, 300 miles away, by early the next morning.

Link 2: The pilot had had a stressful week, with both professional and personal anxiety and bouts of poor sleep.

Link 3: He had gone out with friends the night before the anticipated flight, returning home well after midnight and facing an early wakeup call.

Link 4: The weather throughout the region had been miserable for more than a week, with a stubborn low-pressure system leading to widespread thunderstorms that would pop up unexpectedly and hang on for hours with little movement. Away from the storms, ceilings were well above minimums, but VFR flight was marginal at best.

Link 5: The airplane the pilot was anticipating using on the trip was a model he'd flown only six times before, racking up about twelve hours.

Link 6: The destination airport was served only by a GPS

approach, and the GPS receiver installed on the airplane was a model the pilot had never used to fly an approach.

Nevertheless, the pilot and his family had the bags packed into the car when he called for a final weather briefing and to file an IFR flight plan for the 300 nm trip. Despite an uneasiness that gnawed in the pit of my, er, his stomach, the flight so far was a go. It was then the briefer delivered Link 7, the final blow.

The winds at the destination airport were measured at ninety degrees to the runway at 18 knots, gusting to 25, exceeding the airplane's demonstrated crosswind component.

It finally dawned on me that although the flight I was planning certainly was possible, there were too many red flags popping up in too many different areas to make a safe outcome reasonably certain.

I was reluctant to abort the flight for several reasons. I wanted to fly. It provided the perfect demonstration of the utility of general aviation, trading a 6½-hour drive for a 2½-hour flight into a small town not served by airlines. And those four hours of not being cooped up in a vehicle with three small kids had incalculable value.

But when I added up the plusses and subtracted the minuses, I concluded that the airplane should go back into the hangar. And I'm sure Dr. Joe, the airplane's owner, is glad I did.

REMARKABLE DETAIL

Recently when closing out a logbook and starting a new one, I wandered back through the "remarks" column to see what memories might come forth. I was surprised, however, when it turned out that the majority of my most memorable flights had nothing in that column, or the remarks were limited to instrument approach flown or some other flight minutia.

The reason, of course, was because the flights that stand out most vividly usually involve something I don't really want to reduce to writing. Yeah, these are the flights where you drive away from the airport with the unsettling feeling that you've just dodged disaster.

It's interesting how those disasters evolve as the years pass and the hours accumulate. For example, my first logbook contains an entry in which I log simply "x-wind ldg" as if it's a feat of which to be proud. I had about seventy-five hours total time and had not flown a 152 since getting my private certificate nearly four months earlier.

On this particular flight, however, I was taking up a passenger who had never been in a light plane before. We made the same fifty-one mile trip I'd flown on my first cross-country flight. On landing, with a crosswind from the right, I nearly lost it off the left of the runway. I don't recall how strong the wind actually was, but in my mind's eye I see what I would now consider a mild breeze that would hardly register on the angst meter.

On another early flight, my remarks only include the names of some friends I was taking for a ride. My memory of the flight, however, includes a half-hearted calculation of weight and balance in which I estimated their weights without asking and ordered fuel "to the tabs" but didn't really object when the tanks were fuller than that.

The takeoff roll was longer than normal, but I wasn't used to flying Warriors at max gross, either. Rotation, however, was another matter. Looking back, it's clear either the cg was aft of the limit, or I'd neglected to set the trim properly. Or maybe both. The airplane was probably overloaded, because I recall it flew much better after ten or fifteen gallons of fuel had been burned.

There are other memories in old logbooks as well. Instructors who overwhelmed me with poorly briefed tasks. Those who pencil-whipped flight reviews or instrument checks. The one who absolutely, positively refused to demonstrate stalls in the Mooney I'd just bought.

But lest you think all the memories are bad, there are plenty that bring a smile. My first real instrument approach by myself. The flight home after my faded Mooney got a stunning new paint job. My first trip to Oshkosh for AirVenture. Taking delivery of a new Citabria.

The regs are very specific about what you must log, but beyond that you may log any information you wish. Some people like to keep track of such things as instrument approaches and holds because they illustrate clearly whether you meet legal IFR currency requirements. But after a certain point, the logbook becomes your flying history, rather than simply a legal record. As such, they allow you to travel backward in time and think of your flying experience.

Yeah, then you get into the whole notion of reducing to paper things the authorities might not want to see if an FAA inspector was to, say, ask to see the record of your last flight review. That leads some people to keep two sets of books—one for the feds and one for posterity. Think about it; you may come up with some new ideas about what you ought to log.

WHERE'S THE RISK?

Ask yourself the following question: Which has more influence on the safety of your next flight, inflight weather-related decisions or stick and rudder skills?

Odds are, those who use their airplanes for serious transportation will opt for the weather decisions, while those who do primary instruction or fly recreationally will choose the stick and rudder skills. The accident record shows that both are important, and for all kinds of airplanes. Just as there are Super Decathlons that occasionally stumble VFR into IMC, so there are Cessna 210s that depart the runway following a loss of control.

Good risk management practices dictate that you assess the type of flight you're about to make, your airplane, your skills and familiarity with that airplane, and the weather you'll hit. Basic stuff, right? Now change the kind of airplane you're flying and see what it does to your decision-making strategy.

Once upon a time I found myself re-examining the state of my flying. Our roomy and reliable Piper Lance was decently equipped for IFR flying, with a lightning detector, Garmin 430, and two-axis autopilot. It had redundant vacuum sources, redundant nav/comms, redundant GPS, and speed mods.

The Lance is a fairly simple airplane to fly, forgiving and docile, despite its description by the FAA of being "high performance." Clearly the biggest determinant of how risky a given flight was likely to be was the pilot's decision making.

However, I found that as the years passed we used the airplane less and less as the family-hauler and business-related cross-country machine it is designed to be. My personal flying started leaning decidedly toward recreation rather than transportation.

With that in mind, the Lance went on the sales block and found a home with some good folks in Alabama. Replacing it in the hangar (and in my logbook) is an American Champion Adventure, better known by its popular moniker of Citabria.

Trading a 3,600-pound traveling machine for a 1,750-pound taildragger required quite a shift, both in operating technique and in-flight risk assessment. While I had successfully landed the Lance with winds of 30 gusting to 50 off the side of the airplane, the Citabria generally stayed tucked in its hangar during such weather. (Well, there was that *one* time...) But the Citabria went to work on those gorgeous Florida days when there's no place much to go, a time when the Lance would hardly be calling me away from my desk.

The assumptions and cockpit discipline that go with flying IFR cross-countries are far different than those in a tandem taildragger. So are the mechanical skills and perceptual challenges. When you assess your own risk, keep in mind not only where and how you're flying, but what. The differences may surprise you.

THE PROP SPINS

When I arrived at the airport, the line crew had taken the airplane out of the gang hangar in Smyrna, Tenn., leaving it on the ramp in front of the hangar door. Due to a slight slope in the ramp, there was a chock in front of the nose wheel; otherwise, the airplane was unsecured.

As I did the preflight inspection, I noted the chock and also the fact that the airplane would roll downhill if it were removed. Rather than setting the parking brake and removing the chock, for some reason I decided to defer removing the chock until we were packed up to go.

Preflight completed, I loaded the bags into the back of the Mooney and waited for my three passengers to emerge from the facilities. Finally, toddler and infant were secure in their car seats in the back of the airplane, and my wife and I boarded for a weekend at the beach.

As always, the Mooney fired effortlessly, and I tuned ATIS. As I was recording the information, I noted out of the corner of my eye someone moving toward the leading edge of the left wing.

The FBO manager was motioning that I had left the chock in front of the nose gear and indicated that he would remove it. Before I could kill the engine with the mixture, he bent down within inches of the spinning blades and pulled it out.

He was aware of the prop danger, but I still get the willies when I think of how close he cut it.

Fast forward to another flight. I was leaving an airport for the flight home after talking business with the owner of a flight school. I was on the ramp with the engine running, awaiting a pause in the busy ground-control chatter so I could pick up my IFR clearance.

Again, motion at the leading edge of the left wing caught my attention. The owner of the flight school was approaching the airplane and motioned me to open the small storm window. Before I could shut down the engine, he was standing between the wing and the prop, intending to stuff some logo merchandise through the tiny hole.

That one, too, makes me think of how close some people like to cut it. Those incidents are why I have been a bit shocked by recent aircraft designs that mount the steps to enter the cabin at the leading edge of the wing rather than the trailing edge.

All other things being equal, why in the world would you want passengers routinely stepping between prop and wing? On some small designs, an aft step might put enough weight behind the main gear that the airplane stands on its tail when a heavy person boards the step. To me, that means perhaps the gear should be moved, not the step.

Sometimes even aviation professionals lose sight of how dangerous the prop arc is. But when it comes to a spinning blade, "close" to too close and, in case of a tie, you lose.

OVERLOOKED

Until I let someone else fly my brand-new airplane, I didn't realize what a disservice I was doing to myself.

As the only pilot who had flown the airplane since its delivery flight from the factory, I was quite used to predictable preflight inspections. I wiped the bugs and grime off the airplane after every flight. I flew with a "zero squawks" policy, in part because of the aerobatic nature of the airplane and in part because it didn't have much redundancy. Mostly, however, it was a nice airplane and I wanted to keep it that way. All in all, I was pretty sure what I'd find the next time I opened the hangar door.

Because of that, and a certain confidence instilled by the airplane's recent manufacture, my pre-flights slowly eroded from being thorough exams to somewhat cursory. But I'm getting ahead of myself.

A friend asked if he could use it to entice his kids into taking lessons. I wrung my hands a bit at the notion of handing the keys to someone else, but this guy's a better tailwheel pilot than I am, so I figured it could do no harm.

As it turned out, being afraid of what he'd do to it was the last thing I should have worried about. After his first flight, he called to thank me for the use of the airplane, but then a pregnant pause hung over the conversation.

"But there was something about the airplane I wanted to talk to you about."

Uh-oh. Visions of ground loops and bent wings and broken longerons and scorched tires and ruined props danced momentarily in my head.

"Yeah, uh, I noticed some corrosion on the horizontal stabilizer's strut brace fittings."

To my credit, I responded with a blank, "Huh?" I'm sure he was impressed with my grasp of aircraft dynamics.

The next day, I hightailed it to the hangar and, sure enough, spotted a pair of corroded fittings that I had overlooked on at least a hundred occasions. Although I wasn't planning to fly that day, I conducted perhaps the most thorough preflight of my career and noted a couple of other spots where my diligence has been less than exemplary.

I found a few spots where my cleanings had missed. A missing screw on one of the wheel pants. A chip in the paint on the steel spring landing gear strut.

While I resolved to do better henceforth, I began reflecting on how I had started down the road to complacency, spurred in part by my conviction that a new airplane should be squawk free—without ever considering at what point "new" became "used" and therefore warranted a closer look.

While this lesson was particularly poignant, pre-flights have long been a subject I was interested in as a human factors exercise. Pilots bring their unique biases, habits, and myopia to every preflight they do.

In a way, having several people fly your airplane is a

method to ensure that it gets the best look it can get. While multiple pilots raise other issues that are not so savory, this is one aspect in which most definitely makes merrier.

RED LIGHT

On a flight that promised to be like many others, the airport was washed with the last glimmer of light from the setting sun. I was in the run-up area checking governors, four sets of mags, a couple of alternators, and all of the other stuff that needed to be done before taking runway 32.

The red area light over the left seat was out, although the right one worked. I squinted into the gloom at the printed checklist. If the checklist was hard to read, I couldn't imagine trying to read an approach plate. I moved the checklist all around, tilting it this way and that, in a vain attempt to find enough spare light to let me follow the plastic-covered card. Frustrated, I reached for my flashlight and resigned myself to needing three hands to do what I was about to try.

On the way out, I had casually mentioned to my right seater that I'd had a *day*. He signed knowingly. "Me too." The guy I'd hoped to be my safety net was admitting that he might need one as well.

That conversation still fresh in my mind, I turned to the right seater, removed the flashlight from my mouth, and said, "Okay, one bit of preflight clarification. If you are as beat as I am, we have to ask ourselves if what we're about to do is foolish."

There. It was out there.

Fortunately, we were not embarking on a five-hour leg through nasty weather; we were launching on an instrument proficiency flight. Shoot some approaches and get some night work in to try to sharpen the edge. We talked about it a little and decided we'd launch with a zero tolerance for complications. One factor out of whack, and we'd return and land. Just having the discussion allowed us to focus on the potential hazard, thereby reducing it. More importantly, we modified the night's plan to exclude simulated emergencies, lest we create a real one.

We departed to shoot an approach at a nearby airport, and both the departure frequency and the destination airport's tower frequency were quiet. First, I shot a GPS approach to a downwind runway, marveling at the way the wind helped me get slightly behind even the pokey Seneca. I flew the approach to a landing, and upon taking off again was quickly greeted by black hole effect as we climbed out over the ocean. Say what you want about flying in the clouds, but even after more than twenty years of flying I still find night over water a little unnerving, regardless of how many engines I have or how good the autopilot is.

I got vectors for another GPS approach to a different runway at the same airport, and as I was coming around, it was clear the lack of cockpit illumination was hurting my ability to run checklists and brief the approach properly. My companion tried to help by holding a flashlight, but that was only a partial solution because he could not anticipate where in my field of view I wanted light. As

I passed the final approach fix for a second landing, we decided our efforts might better play another day. Another landing—this one with a substantial crosswind—and we requested vectors for the ILS at the home field.

That approach went smoothly, and as I was taxiing in I was rather surprised at how tired I was for having made only a 1½-hour flight and three instrument approaches in relatively benign weather. Before the flight we'd joked about the "IMSAFE" acronym, and each of us declared fitness for flight, despite the fatigue we'd both admitted to.

Even in retrospect, I was not convinced it should have been necessary to scrub the flight, though I was glad we'd agreed to forego single-engine work or other "trouble." But the flight served to show once again how being conscientious isn't always enough. Sometimes you have to actively look for trouble, and even then you might not find it.

The old saying that it's better to sweat in training than bleed in war is nowhere as clear as it is in flying airplanes. With each layer of complexity—engines, avionics, systems, airspace, malfunctions—the quantity of sweat increases. Most of us have had enough trouble-free hours of flying that we forget it might not be that way someday.

Plan to sweat. Make the effort to refill the vessel that will get drained when the chips are down. The elusive goal: to always be able to answer "yes" when circumstances steal a line from Gatorade and ask, Is it in you?

WHEN "BROKEN" ISN'T

Airplanes are not now, nor will they ever be, like Toyotas. They are not jump-in-and-go, just-put-gas-in-it machines. An FSDO friend of mine likes to say he's never seen an airplane that met the full legal definition of "airworthy" in every respect. In virtually every airplane out there, he contends, there lurk gremlins, wear points or omissions that violate the letter of that complex pile of paper we call the FARs.

Taken to the proper extreme, perhaps he is right, perhaps not. But I believe it is safe to say that inside the vast majority of airframes lay a few quirks that might stick in the craw of someone sufficiently motivated to find fault. However, in some sizeable fraction of those airplanes so affected, the bugs rise to a level above idiosyncrasies and into the range of squawks.

With a show of hands, who has never flown an airplane with a known squawk? I thought so.

Among the vast range of judgment calls a pilot must make is when "not quite right" becomes "not acceptable." And I'm not talking cord showing through the tire tread or fuel dripping from wing bladders here. I'm talking about that right brake that's a little spongy but comes right up when you pump it once or twice. That left mag on the left engine that runs just a little rough when you do the mag check. That autopilot that lets the airplane wander back and forth in heading mode or hates to capture nav two.

Your airplane has two of a lot of stuff, but often there's only one pilot. Even two pilots is no guarantee.

I recently heard a story from a pilot training for his multi in a Seneca II. On initial climb, one engine started delivering only partial power, with a commensurate drop in fuel flow. The student pilot initiated a turn back into the pattern to land, while the instructor "fiddled" with the engine controls. The instructor found he could restore power by leaning the engine somewhat, and the pair continued on their way across one of the Great Lakes as night was falling.

To some pilots, this is a shocking case of get-home-itis, leading pilots to ignore a known deficiency. Surely there have been enough accidents caused by this kind of dynamic. To other pilots, the now-smooth engine, plus the one on the other wing, would be enough to make the earlier hiccup a non-issue.

The fact is, no matter which side of the fence you might come down on in this case, it's the same kind of decision pilots must make routinely if they hope to get any kind of reasonable dispatch reliability out of their airplanes.

Lest you think this is a tirade against aircraft reliability, it isn't. It's quite clear that even airplane designs with a history of being problematic can be made reliable through aggressive maintenance plans and meticulous attention to detail. That said, there will always be those issues that make you stop and scratch your head. Even if you have a published, approved minimum equipment list, there is enough gray area between "working" and "broken" that many items installed on the airplane can be only somewhat working. And still pilots will fly it at night in instrument conditions over the mountains.

How, then, do *you* decide when close is good enough? Certainly short hops in benign weather lead to different conclusions about "airworthiness" than do long jaunts, even if literal interpretation of the regs allows no such thing.

Pilots make a great deal of noise about setting personal minimums and sticking to them in the face of temptation to push the envelope. That works with, say, weather because the weather is quantified; you know the crosswind is X and the ceiling is Y. But when is a mag check rough enough to set off the alarm? There's the drop, sure, but what about a little choppiness when the drop isn't excessive? Gray, gray, gray.

Over time, pilots learn little tricks, habits, and excuses for making these kind of calls. But just because you've been doing it for umpteen years doesn't mean you've been doing it right. And is there even a "right" way? Truth be told, that's the kind of wisdom we could all use a little more of.

THE SAFETY STOOL

The voice on the other end of the phone was insistent. "The very fact that you felt you had to call me to ask my opinion of the weather demonstrates that you should stay on the ground."

I was talking with a veteran flight instructor about the willingness of some pilots to ignore the responsibilities of flying for long periods of time, and those same pilots then complaining or making excuses when their proficiency is not up to snuff.

He told me of a call he once got from a former student who called him about the crosswind at the airport the pilot was trying to depart. It was 20 knots across the runway, and the pilot wanted an opinion on whether to try to take off. That line above was what he delivered to the uncertain pilot.

I, for one, admit I'm a weather wimp. I tend to be loaded to not go unless, as the Magic 8 Ball sometimes says, "all signs point to yes." Sure, there are times when I depart despite a nagging doubt, but I try not to assume I'm taking off every time I open the hangar door. And then there are the times during preflight planning when you find yourself out of instrument currency or discover the ELT batteries are due.

Although the FAA sets minimums for recent experience, it's an old saying that what's legal isn't necessarily safe. Three takeoffs and landings in three months is not enough, particularly if the number of pages in your logbook doesn't exceed your IQ. As the speed and sophistication of the airplane increase, so too should your motivation to maintain proficiency.

Lots of pilots use general aviation for business, at least some of the time. But just as business affords us the motivation to fly, it can sometimes saddle us with responsibilities that prevent us from flying. And the yellow light should go on when the days start piling up since last flight (or last approach or last hold or last night landing or, well, you get the idea).

This quest for proficiency is the rationale behind the annual or even semi-annual recurrent training a lot of pilots routinely undergo. Whether required to by their insurance or simply inspired to by their conscience, pilots who train frequently tend to be safer pilots.

No surprise there, but consider also that recurrent training of the commercial simulator company variety cannot completely prepare you for real-world situations in which you may find yourself. They can do a lot, and with such things as system malfunctions, the real value may lie in teaching you to remain calm and fly the airplane while troubleshooting the problem and developing a coping strategy.

But what simulators can't do is replicate the more human side of the experience. Will you go back for another try after a missed approach rather than fly to your alternate when you know you absolutely, positively have to get on the ground to make an important appointment? Will your eagerness for that vacation destination tempt you to eat into your fuel reserves when the headwind is more than forecast? Will you be the one saying that a crosswind gust came out of nowhere because you've been too busy to keep up your crosswind stick and rudder skills?

Safety in flight depends on three legs: solid equipment, stick and rudder skills, and pilot decision-making. Being human, we all have a tendency to trust that two and a half legs are as strong as three. Many times they are, but remember you are betting your life on it.

If this sounds all doom and gloom, it shouldn't. If there is any positive borne out of the fact that aviation is such an expensive undertaking, it's that the people who operate high-end airplanes are, for the most part, competent and conscientious in their affairs because that's what allows them to absorb the cost of flying in the first place. So from that standpoint, much of this might be preaching to the choir. If you're singing along, good for you.

The history of aviation is littered with pilots who bought more airplane than their skills could handle, or who found the hard way that superior skills won't necessarily outweigh poor judgment, or who hoped their piloting chops would save them from an unairworthy airplane. The erosion of skills and conflicts on judgment are insidious forces that erode flight safety. Couple that with the fact that the fleet isn't getting any younger, and it might be time to consult your songbook again, even if you've been in the choir a long time.

DIY, SORT OF

Ever think you might be screaming through the air at more than 300 knots in something you built yourself?

When the FAA created the "experimental amateur-built" certification category in 1952, the concept was aimed at tinkerers who aimed to build a simple, low-and-slow aircraft just for the heck of it. Indeed, at the time instrument ratings among non-professional pilots were by far the exception. The face of general aviation looked entirely different then.

Fast-forward more than half a century, and you could put money down on a single engine jet with a cruise speed of more than 320 knots, and yet is as simple to fly as a Bonanza. You will have the privilege of building this airplane yourself, and afterward can do your own maintenance if you so desire. The upside is you won't have to spend ten years doing it. Instead, you can work at Epic Aircraft's facility in Bend, Ore., and have it flying in a year. You, yourself, will spend perhaps ten to twelve intensive weeks, while the rest will be done by hired guns.

By now you're wondering how this is legal. You're asking whether such an aircraft can possibly be safe. Most of all, you're afraid that if *you* tried it, the FAA would swoop in and declare the project a $1 million lawn ornament.

Amateur-built aircraft now make up about twenty percent of the general aviation fleet, and a fair amount of the innovation that has trickled into modern aircraft has gotten its start in the experimental field. Heck, even Cirrus Design is rooted in the Klapmeier brothers' early foray into designing and building their own airplane. The FAA has some measure of experience in the homebuilt arena.

The common misconception is that, in order to certify an airplane as amateur-built, you have to personally build half the airplane. That is flat-out wrong. At what point does a component become a part? Are you expected to build the engine? Mill the metal to make the parts to build the engine? Mine the ore to make the metal to mill into parts to make the engine? Of course not. That would be ridiculous.

Instead the FAA has defined a group of tasks involved with building the airplane, more than half of which you must accomplish. And here's where builder assist comes in. You don't have to wire *all* the circuit breakers; you can just do one and then hire someone to do the rest. That's the basis on which builder assist programs are founded. And legally so.

However, those programs have been abused by "serial builders" who hire themselves out to build kit planes for others, in clear violation of the spirit of the so-called fifty-one percent rule. The FAA caught wind of it, of course, and launched into an examination of how the fifty-one percent rule is applied by different Flight Standards District Offices when a builder comes in for an airworthiness certificate.

The FAA said it would not re-examine any kit programs it had already deemed met the fifty-one percent rule, and would not alter any existing airworthiness determinations. However, when it comes to regulating how much professional help a builder can have, all bets are off.

The upshot is that there are a few handfuls of pilots out there flying around in pressurized cabin-class turboprops they built themselves, with a little help from the factory. Do they know something the rest of us don't? In talking to a few of the builders of the Epic LT, I was struck by a couple of things. First was just how seriously they take the construction project. Flying at FL280 at 300 knots is not something to take lightly in any respect. But in looking further, I noted how the builders, many of whom are supremely qualified as both pilots and engineers, make improvements on the original design that are then shared with other builders. In a certified aircraft, of course, this is a much more difficult proposition, which adds cost as well as discourages minor fixes that the manufacturer doesn't deem important enough to warrant correcting.

I was pleased to hear two Epic LT builders talking about ways to improve the emergency oxygen setup and how to make the battery easier to service. And because they're all building their airplanes in the same place, and with factory support, they can look over each others' modifications and run them past company engineers to see if the change might lead to any unfortunate side effects.

In a lot of ways, that process makes sense. And to then have the ability to fly those airplanes hundreds of hours each, modifying them almost at will, before launching a certification effort means those pilots who want to wait for a store-bought airplane get the benefit of that wisdom.

So would I fly 300 knots in something I built myself? With this kind of support in place, you bet I would.

INQUISITIVE CULTURE

On a day like so many since, a crowd had gathered at Ft. Myer, Va., to watch an airplane. On Sept. 17, 1908, airplanes were novelties seen by few and piloted by even fewer.

On this day, Orville Wright and Army Lt. Thomas Selfridge took to the air over the parade grounds on a demonstration flight for the U.S. Army Signal Corps. About five minutes after takeoff, while maneuvering about one hundred feet above the ground, the nose dropped and the airplane crashed. Selfridge was killed, the first fatality in the nascent world of powered flight.

While there would be many more over the ensuing century, this accident was noteworthy because it set the tone for what has become *de rigeur* in aviation: the deconstruction of the accident to examine why it happened so operators can take steps to make a repeat occurrence a little less likely.

In the case of Orville's crash, the crash investigation involved Orville and his mechanic looking for some kind of mechanical issue.

Given the technology of the day, a mechanical issue would not have been surprising, and the accident was eventually pinned on a propeller blade striking one of the wires supporting the rudder. These days, investigators would also consider the possibility of human error, and in this case a stall/spin would seem to be one of the usual suspects.

A few days after the accident, *The New York Times* stated, somewhat presciently, "The aeroplane is still far within the experimental stage. The perfected machine will doubtlessly be different from it in everything from principle to motive power."

Whatever the true cause of the Wright crash, it set in motion the tendency, in fact the *need*, to examine accidents. Boating magazines don't examine accidents, nor do motorcycle magazines or just about any other pursuit you might wish to examine. What makes airplanes different?

The National Transportation Safety Board, itself more than forty years old, is charged with investigating aviation accidents, although it does rely on FAA inspectors to report on many minor ones. Within a few days of an accident, the NTSB will publish a preliminary report, available in a searchable database on its web site. A year or so later, a report giving probable cause is generally released, although a small minority of accidents takes much longer.

Aircraft shoppers often use the database to search for safety issues that might be present in the type aircraft they are considering. Some pilots look up accidents at particular airports they are thinking of visiting or basing their aircraft.

Spend any time wading through the database, which contains nearly half a century of data, and it gets depressingly clear just how common certain types of general aviation accidents actually are. Unlike in the airline industry, where "common causes" has essentially become an oxymoron, general aviation sees a continual parade of accidents that are seemingly interchangeable except for the details of who, where, and when. More than three-fourths of GA accidents are pinned on pilot error, and that number might actually be higher given the number of accident reports that include the phrase "for undetermined reasons."

Now, this should in no way be construed as a call for the FAA to step in with expanded requirements for flight data recorders, cockpit voice recorders, and/or recurrent training. Rather, pilots must pay more heed to the fact that it is *their* behinds that are in the air, and it is *their* responsibility to ensure the safety of their passengers first, themselves second, and the airplane third.

As the logbook pages pile up, complacency comes easy. That's understandable, since any venture practiced enough becomes second nature. But from time to time it helps to step back and put the risk into perspective. Flying several miles above the surface of the earth puts you in a hostile environment. The equipment can protect us from the elements, but it cannot protect us from irresponsibility, stupidity, or ignorance.

That Orville Wright was involved in the first airplane fatality is, perhaps, not surprising. Consider that he was a self-taught pilot, flying a primitive aircraft at a time when both the knowledge of aerodynamics and the enabling technology were in their infancy. In a way, it's somewhat surprising it took five years of powered flight for the first fatality to occur.

In most respects we've come a long way since then, but sometimes, when you examine the accident reports month after month, it seems maybe we've not come that far after all.

RIGHT TURN, LEFT TURN

There are two kinds of people in the world: those who climb into a cabin-class airplane and turn right, and those who turn left.

In a typical cabin-class airplane, the door is in front of the wing, and one ascends the airstair and then either goes into the cabin or the cockpit. The details vary a little from airplane to airplane, but you get the idea.

Some airplanes are made for the pilot; some are made for the passenger. That's not to say there are pilot's airplanes and passenger's airplanes—which there are, of course—but rather that you can tell who the manufacturer had in mind when the designers put pencil to paper.

Regardless of where they plan to sit, any potential owner will look at performance, maintenance costs, mission suitability, parts availability, and operating costs. It's at that point that two kinds of buyers emerge.

A buyer looking for an aerial limousine will then look at cabin appointments, amenities, style, and comfort. Perhaps an employee pilot will check out the flight deck, but the basic analysis is whether the airplane meets the mission requirements, financial objectives, and comfort test.

Other buyers are pilots first. Perhaps they make a cursory glance at the cabin as they make their way to the front office, perhaps not. To this class of owners, such issues as performance, handling, cockpit ergonomics, and installed avionics become paramount.

And then there are the couples who tag-team an airplane. At the risk of sounding sexist, these buyers typically consist of a male pilot who turns left, and a female passenger who turns right. Cockpit and cabin. Nowhere to hide.

Although I personally fit firmly into one of the above categories, I'm not denigrating or criticizing anyone who puts priorities in a different place. After all, the variety of airplanes on the market represents the opportunity to tailor your hardware to your own personal preferences. And who could possibly argue with that?

The one thing I think we could all agree on would be the need to keep operating costs in line. Whether your direct operating costs are $400 an hour or $2,000, it remains crucial to keep costs in line to better enable you to justify owning the airplane in the first place. Sure, the emotional and practical value of having the airplane at your disposal is worth something, but it doesn't allow you to completely turn a blind eye toward costs.

For instance, a couple I know once bought a Baron they used extensively for personal travel. However, a series of maintenance events and some business setbacks put them in a hole that led them to drive on 500-mile trips, the very kind of trips for which they bought the Baron in the first place. Another fellow was looking at moving from a high-performance single into a light twin. After crunching the numbers, he realized it would make better financial sense to keep flying the single and, when he needed the added capability, charter a locally-available King Air.

There is a difference between "ought to" and "will," and therein lies the dilemma for many pilots. At the low end of the airplane market, there is a rule of thumb to buy the airplane that will work for 85 percent of your missions and rent for the other 15 percent. Renting—or chartering—however, gets a bit problematic as the missions become more specialized. And that leads many people to buy far more airplanes than they "need" on a routine basis just so they cover the bases on those once-in-a-while missions.

I say, if you can afford it, then do it. After all, if you have more seats than you need, ultimately you will find some folks to fill them from time to time, which helps expose newbies to the wonderful world of general aviation. And right now we need all the friends we can get.

However, if you do take this route, do it with common sense. Don't overextend yourself for the sake of your ego. Don't take shortcuts on maintenance, training, or operations for financial reasons. Always keep in mind that, no matter which way you turn when you enter the cabin, your first priority should always be the safety of the flight.

But you already knew that, didn't you?

PART IV
THE POLITICS OF FLYING

SETTING THE TONE

Go ahead. Make an instructor's day. Volumes have been written about the state of flight instruction. Veteran pilots sometimes chafe at the notion of paying some kid barely out of puberty to approve them for instrument flight for another six months or, worse yet, sign off on their ability to fly at all for another two years. New pilots wonder what they're getting when their instructor clocks in with a minimum of experience.

The notions get worse as even those instructors vanish in a few months, snapped up for a lousy job by a tiny commuter airline and replaced by someone even younger and with even less experience. Their most redeeming quality is that they're cheap to hire.

Then there are the part-time instructors. They have other jobs and instruct, not to build hours, but because of the rewards it brings them. You get the feeling that the income derived from instructing is nice, but it's not the reason they're there. They can be great instructors, but they tend to be hard to book because of their non-instructing career. They generally cost more, but they're often worth it.

Finally, you have career instructors with lots of experience who don't give a damn about how many hours they're accumulating. They're often the most expensive of the lot. They tend to be self-employed and figure that someone who's paying a mechanic $65 an hour to work on the plane ought to pay a reasonable fee to learn to fly the plane.

The economics of instructing force many people in all three categories to give it up. But there is something all pilots can do to encourage quality instructors—and it doesn't cost anything.

Try.

Who can blame an instructor for getting disillusioned by cheapskates who treat everyone like a waiter and know-it-alls who feel like the instructor's signature on a flight review is their birthright? Amid the disillusioning customers are the ones who really want to learn, who really do their best, and who put a lot of effort into getting it right.

Ask any instructor what that means to them, and odds are not a single one will say, "Another $30 in my pocket."

By trying hard to achieve the kind of perfection that eludes everyone, you show your instructor that he or she is making a difference. That means more job satisfaction, which might keep them instructing longer, which helps everyone who comes after you.

I hear a lot of veterans say that flight training today is minting worse pilots than it did years ago. To them I say, take a look at the economics of instructing. Is it any wonder good instructors are hard to find?

Personally, I demand more of the instructors I fly with now than I did in the past. As a result, I have stopped throwing my money at revolving-door instructors and opt instead for experienced instructors who take a professional approach.

If all pilots did that, maybe more good pilots would consider instructing as a long-term option. And that would increase safety for us all.

NO GRAY HAIR HERE

The sad state of flight instruction is nothing new to any general aviation pilot who has tried to find someone good when it came time to upgrade a rating, get a flight review, or take an instrument check.

Think you can make it all the way through a license or rating without losing at least one instructor to the call of higher pay at a commuter airline? Bet ya can't.

For a few lucky years, my CFII of choice was an anomaly—a dedicated professional who was a skilled veteran with a love of teaching, not just a kid looking to build time before jumping to the airlines.

That all changed when my CFI (sounds as possessive as a high school romance, doesn't it?) went and became a designated pilot examiner. For the flying public as a whole, this was good news. The standards he holds are high, and the applicants will have to know their stuff to get past his book of pink slips. Personally, however, it was a devastating loss.

As I scoured the area for a veteran instructor, I came up empty for the longest time. Judging from the comments I got from others to whom I complained, I know a lot of other pilots were in the same boat.

The path of least resistance is to throw your hands up in disgust and reconcile yourself to the fact that you're going to have to hire some kid whose flight time is barely higher than a starlet's IQ to sign off on your abilities for the next two years or six months or whatever. While you might learn something, it's more likely you'll either teach the CFI a couple of new tricks, or you'll hear his or her headset tap-tap-tapping against the right-side window and look over to see him or her catching up on some sleep.

I recall on one occasion checking out a rental Mooney 201 at a local flight school. I wanted to have a backup airplane in case my Lance was down and, as an ex-M20J owner, I thought this might fill the bill. During the preflight, I taught the CFI a few things about assessing the condition of the rubber biscuit shock absorbers and draining the Mooney's static system. I convinced him the thing takes off better with partial flaps and showed him how to stop bouncing his landings.

When we got back, I paid for his time, and I paid for the airplane's time. Good for me.

As pilots, we get used to writing out checks. I don't know of anyone who minds paying a bit for some knowledge that may someday make the difference between a safe landing and a long way home. However, we also have to accept that we must all too frequently compensate others for the privilege of performing for them. But that doesn't mean we have to like it.

REPERCUSSIONS

The worst thing anyone could do with an airplane has now been done. Several times.

For more than a generation, the common practice at the airlines was to cooperate with hijackers, and in the end everyone would go home. Then a gang of suicide hijackers changed the rules for everyone.

After the initial burst of panic, the dust settled and flying became rather more like before than less so, with some important differences for those who fly around the nation's capital and wherever the president happens to be.

However, the general public and the government have since cast a wary eye at pilots who could "fly wherever they want to with no one knowing"—despite the fact that drivers have relatively more freedom in that respect, though perhaps not the speed. Because of that, and because perception is often more important than reality, pilots were forced to begin acting a little more worldly.

Being more of a grownup about being a pilot means keeping yourself, your equipment, and your performance above reproach. The checklist I like:

- Secure your airplane. Make sure the cabin door remains locked when you're not around. If your lock is suspect, as most are, invest in a throttle lock or prop lock to help prevent your airplane from being stolen and used for mischief.

- Keep your skills as proficient as possible. Airplanes that fly by the numbers on assigned headings are less likely to be perceived as threats than airplanes making low passes over the Ferris wheel at the county fair or flying haphazardly over reservoirs at low altitude. The dividends of such an approach will serve general aviation well and carry the fringe benefit of increasing how safe a pilot you are.

- Cooperate with airport security. Many of us take for granted the ability to drive through a gate, cross taxiways and ramps, and park next to our T-hangar or tie down. At many airports, ramp access has become limited or even eliminated. Figuring out ways to defeat the security requirements makes life easier in the short run, but what's it say about the trustworthiness of pilots in general?

- Be a good neighbor. If you see someone sneaking around, don't just shake your head sadly and drive away. The airplane you save may be your own.

- Know who you're flying with. In a small fraternity like aviation, it's easy to take everyone at face value and accept the bond shared by pilots, even without much proof of who you're dealing with. While an offer to go flying remains one of the surest ways to make friends of an airport stranger, practice a little extra caution.

Copycat psychos and other would-be martyrs have seen what an airplane can do. Fortunately, general aviation aircraft did not become the tools of choice for later hatemongers, but that doesn't mean it'll never happen. There's no reason to make it any easier for them than it has to be.

TAROT SECURITY AGENCY?

As a nation, we have become so paranoid about aviation security that a self-described psychic got a flight canceled because it may have contained a bomb. What's this got to do with general aviation? Plenty.

Ever since that *day* in September, general aviation has been a whipping boy of those who would have us be secure in all of our effects. Not the only whipping boy, to be sure, but a substantial one. This despite the lack of any kind of publicly revealed evidence that small airplanes represent a substantial threat to any building, landmark, congregation of people, presidential motorcade, or air convoy.

But then a "psychic" induced the Transportation Security Administration to let loose the bomb-sniffing dogs, ultimately causing cancellation of a commercial airline flight because the search delay put crewmembers past their duty limits.

If there was any justice in the world, the bureaucrat who made that call would have been exiled to passenger screening at the terminal in Reidsville, Ga. (Look it up.) But of course, that didn't happen. Instead, the incident served to further chip away at the public's ability to look at airplanes as anything other than noisy threats.

For general aviation pilots, this episode added to the wake-up call that's been ringing on the bedside table for nearly a decade. People don't trust airplanes, but airliners are too useful to abandon, so let's let the other guys have it. However, instead of crying in our corn flakes, it's past time we took matters into our own hands and went on the public relations offensive.

Don't befuddle your neighbors with statistics on general aviation's economic impact. It isn't going to work. Instead, do what you can to illustrate on a visceral level why it is we put up with flight reviews and maintenance bills that would make a Mercedes dealership squeal with glee.

Don't run afoul of any regs on common carriage or private certificate reimbursements, but do what you can to show people what airplanes are good for. That means a Sunday pancake run, a business meeting three hundred miles away, or a weekend getaway.

This battle is grass-roots all the way. Let the alphabet groups worry about lobbying the elected guys, but without a greater public appreciation for small airplanes a lot more airports will suffer the fate of Meigs.

Despite the relatively clean-cut ranks of pilots, we are in many ways faced with staging a protest movement. A protest against encroaching regulation, skyrocketing costs, and outright prohibitions on flying where we cannot possibly do any harm. This won't happen today. And it won't happen next year.

This is a continuing effort to help non-pilots understand that we're just like them. They don't need teams of armed guards protecting them from us. They don't need metal detectors or triple-layer concertina wire.

In fact, they don't even need a psychic. Bet you knew I was going to say that.

ARRESTING THE JAILER

As pilots we hold ourselves to a certain standard of proficiency. The aircraft we fly are maintained above a threshold of repair. Everything aviation related is measured against a benchmark, and woe unto he who comes up short.

Cruising in the flight levels at more than 200 knots or shooting an approach in poor weather in the dead of night are hardly places anyone wants to be when their skill or equipment is found wanting. Just to make the stakes even more interesting, the FAA is always there, looking over your shoulder, holding the power to make your life miserable should any shortcomings appear.

Why, then, is the FAA itself not held to the same standards?

Case in point was the conversion from FAA-run Flight Service Stations to Lockheed-Martin's contracted-out, uh, product. The stories are legion, and to that pile I add one of my own.

I was flying from Jacksonville to Fort Lauderdale one morning, and, like an idiot, I had missed my routine of getting a briefing and filing my flight plan by DUAT the night before, which I would then update just before departure. I was under a little time pressure, and so rather than boot the computer I decided to file from the car while driving to the airport.

Big mistake.

I got what had become the obligatory twenty-minute hold, during which time I realized I had left my headset at home, so I returned to get it. Still on hold, I started back to the airport. Finally, a briefer came on the line.

He apologized and said he was "on a backup of a backup system" and he wasn't sure what he could do for me, but he would try. Okay. I gave him my route and asked for a standard briefing. I heard a few keystrokes, and he said, "Now, the information I'm about to give you is not for navigation. It says that in big red letters here."

As he switched from link to link on his computer, he told me stories about his life in the military living in Jacksonville, but that was a while ago, and he's heard it's grown a lot since then.

Can we get back to the briefing?

The weather was relatively benign, but my departure airport was IMC and scheduled to remain so for the next hour or two. There was no heavy fog, just low ceilings and low visibility. Once above the layer, it looked like smooth sailing, although he couldn't give me winds aloft.

Then he gave me this gem: "Any Notams I could give you would be unreliable. I suggest you check them before you leave." By now, of course, I was at the hangar waiting to get off the phone to open the door.

"I suppose you want to file a flight plan?" he asked.

"Yes I do."

"Well, I'm not sure if it will take it. You'd probably be better off to air file."

Did I mention the departure airport was IMC? The fog and ceilings weren't horrible, but there was no way I was going to take off VFR.

As it turned out, the computer did accept my flight plan, and I hung up and prepared to fly. Since I figured I'd have a few minutes before my clearance would make it into the system, I used the time to review the XM weather on my yoke-mounted Garmin 496, with which I gave myself a DUAT-style briefing to confirm what the FSS had given me.

Have we come that far, that a non-IFR-approved handheld GPS/weather receiver does a better job of supplying weather briefings than the official FAA-sanctioned preflight briefing source? Oops, make that the *unofficial* FAA-sanctioned preflight briefing source, I suppose.

The FAA, in response to Lockheed Martin's startup issues, denied certain performance bonuses, but otherwise resisted any public action that was in any way proportionate to the risk, both legally and safety-oriented, that pilots were exposed to when they used the FAA's primary "legal" briefing system. Do you think the FAA would tolerate such a lack of diligence on the part of the pilot or aircraft owner?

As a postscript, at the end of the day I was checking my email, and I got this from an AOPA email:

Software Update Temporarily Takes Down Flight Service

Pilots had a difficult time getting through to a briefer or filing a flight plan Thursday morning after Lockheed Martin updated software throughout the system. Lockheed officials told AOPA that the software had worked fine in tests but failed under the load of pilots calling for their morning briefings.

By midday the system was mostly restored. *If you encounter a problem with flight service, log your complaint on the flight service comment line* 888/FLT-SRVC.

We pilots hold ourselves and our equipment above reproach, thanks in large part to the combination of the safety carrot and the FAA stick. Too bad the other guys don't seem to share our conscientious attitude.

INSURANCE RULES

Who rules general aviation?

Ask a hundred pilots, and you'll probably hear: the FAA. Ask a hundred operators and you're most likely to get: the insurance companies. For better or worse, insurers are dictating more and more of the terms under which people own aircraft as well as how they use those aircraft.

Bluster about it if you want, but upon closer inspection, it really shouldn't be all that surprising. After all, aviation is a relatively unforgiving avocation, and the insurers are the ones picking up the tab if you should happen to throw a gutter ball. Don't like it? Pay cash for your airplane or finance it in such a way that the loan is guaranteed by something else.

Because the federal government doesn't require aircraft owners to carry insurance, as most states do for automobile owners, you're free to go naked if you want. However, as long as a creditor has a stake in the airplane, that creditor has the right to require you to protect that interest.

Getting to the truth is complicated, because insurance companies suffer many losses that do not appear in the NTSB's accident database. So while an analysis of the accident record may paint a relatively rosy picture of twin operations vs. single operations, the underwriter's payout record might be bleak. Unfortunately, there's no way to independently verify the insurers' position.

Be that as it may, the fact remains that insurers take a dim view of multi-engine operations by non-professional owner-pilots. With the help of a friendly insurance broker, I secured some numbers that illustrate just how significantly insurance companies are steering pilots from twins to high-performance singles.

I got quotes on two aircraft: a $90,000 1964 Beech B-55 Baron and a $320,000 1988 Piper Malibu. Each of these aircraft is suitable for single pilot ops and relatively serious transportation. And each serves as a gateway toward bigger and more capable airplanes. Each has had a period of high losses that were addressed by tailoring the transition training to the airplane. For the sake of discussion, I assumed a pilot with substantial experience in retractable-gear singles but a new multi-engine rating, with twenty-five hours in a Piper Seneca.

The insurance companies came back with similar requirements. First among them was initial training through a recognized program, such as those offered by SimCom, FlightSafety etc. The Baron would require twenty-five hours dual after the training program; the Malibu would require ten hours.

But the differences were eye-opening. The Baron would carry an in-motion hull deductible of $5,000, while the Malibu would be $1,000. And despite the huge disparity in hull value, the premium was actually higher for the Baron than the Malibu: $6,000 per year and $5,800, respectively.

The Baron owner would pay 6.7 percent of hull value every year just for insurance; the Malibu owner would pay only 1.8 percent. Is a Baron four times as likely to suffer a "loss" as a Malibu? The underwriters appear to think so.

How significant is the fact that the pilot in this example was a rookie in twins? From my perspective, I took that position because, well, every multi-engine pilot has to start somewhere, and a B-55 seemed like a good place to start. The other reason is that it's practical to assume the typical pilot/owner of a cabin-class airplane would gain experience through several years of practical airplane use in singles and then migrate to a twin, rather than a person who took the CFI/CFI-ME time-building route.

The insurers counter that they are simply responding to actuarial data and pricing their product accordingly. That may be, but regardless of the rationale applied by the insurer, the net result to the pilot community is the same: a high-performance single is generally cheaper to insure than a relatively comparable twin.

I didn't pursue these quotes through the logical progression, say, by comparing a Pilatus PC-12 to a King Air 200 operated by a highly experienced pilot, but the influence of the insurance companies on the industry is evident regardless of how that relationship would play out. After all, if the insurance companies discourage entry-level twin ownership, then by definition that alters the ranks of experienced twin pilots a few years down the road.

SHADES OF GRAY

A friend of mine was having problems with corrosion on the magnesium inlet of the PT6 on his Piper Meridian. The fix was expensive and temporary, but the guys who sign the aircraft maintenance logs say it had to be done for the airplane to be "airworthy."

Another friend of mine, an FAA airworthiness inspector for many years, claims he has never seen an airplane that was fully airworthy in all respects, both mechanically and legally. The murky areas, then, become how tolerant you are of faults, the extent of the fault, the type of operation, and other issues that look toward setting a practical definition of airworthy, rather than a cookie-cutter one.

I think there are a few points to bear dwelling on. The first is the fact that two different inspectors may come to differing opinions about the airworthiness of a part or installation, even if each professes to have no vested interest in it.

When is a tire "too worn" or an oil seal "too leaky"? When does a slightly loose hinge or a working rivet turn from a characteristic of a used airplane to an airworthiness item? Subjective judgments—sometimes based on years of experience and other times grounded only on "that looks like a problem"—are part of the job for all mechanics. And maintenance shops are torn among conflicting needs to retain customers, replace parts pre-emptively before failure, increase revenue, build a fair reputation, and protect themselves from liability. Some walk this tightrope admirably; others are less successful.

Some parts, of course, have a specific threshold at which "airworthiness" is implied by regulation or by the manufacturer based on time in service, thickness of metal, or some such measurable, objective criteria. But even that is not foolproof, as localized wear, corrosion, or cracks can obviate the broader criteria.

The other issue, of course, is the legal one. While it's true that maintenance isn't complete until the paperwork is inserted in the logbook, the sad truth is that that's not enough anymore, either. Two words on that front: unapproved parts.

Airplanes operate in an unforgiving environment, it's true. But the FAA's regulatory environment, I would argue, does as much to promote unairworthy airplanes as it does to enhance safety.

Excessive regulation is one of the factors that has raised the price of parts to ludicrous levels, which encourages owners to try to squeak out a few more hours rather than address an issue before it becomes a problem. The sad fact is that it's getting ever-more difficult to adopt a strategy of fixing things before they break.

In addition, quality control at some parts manufacturers remain suspect, and failure of new parts in the first few hours in service remains unacceptably high. I believe this is due in part to the manufacturers trying to keep costs down, but I also believe the FAA shares the blame by effectively stifling innovation.

Cutting-edge innovation by definition introduces a measure of unreliability into the final product, and the FAA has attacked that issue with a vengeance. The unfortunate side effect, though, is that incremental improvement is strangled by certification requirements that often bear little relevance to the risk.

No one wants to see auto parts from Joe's Discount Barn installed on a critical system aboard a high-performance aircraft. But our zeal to prevent that problem has led us too far down the path of verification, which leaves us with $60,000 piston engines that rely on technology that predates the oldest active pilots.

For pilots struggling to maintain thirty-year-old aircraft that simply can't be replaced because no company is making them, this problem can pop up on any preflight inspection. The net result, in terms of cost as well as dispatch reliability, is that for some operators it simply becomes more hassle than it's worth.

I believe it is time to think about a different approach—one that relies on modern engineering standards more than regulatory oversight. I do not think the totally self-certifying process used in the Light Sport aircraft category is appropriate for jets, turboprops, or any pressurized airplane. But I think there are elements of that approach that would pay dividends in keeping costs down, improving quality, and yes, even enhancing safety.

Federal bureaucracy being what it is, I expect the status quo for some time. But that doesn't mean we can't all hope and work for a brighter future.

listen|imagine|view|experience

AUDIO BOOK DOWNLOAD INCLUDED WITH THIS BOOK!

In your hands you hold a complete digital entertainment package. Besides purchasing the paper version of this book, this book includes a free download of the audio version of this book. Simply use the code listed below when visiting our website. Once downloaded to your computer, you can listen to the book through your computer's speakers, burn it to an audio CD or save the file to your portable music device (such as Apple's popular iPod) and listen on the go!

How to get your free audio book digital download:

1. Visit www.tatepublishing.com and click on the e|LIVE logo on the home page.
2. Enter the following coupon code:
   d144-afeb-0d81-715b-2194-4b8f-28cc-9d9c
3. Download the audio book from your e|LIVE digital locker and begin enjoying your new digital entertainment package today!